Advances in Sewn Product Technology

The fashion industry continues to contribute significantly to greenhouse gas emissions. It is one of the biggest polluters, one of the most wasteful of all global industries and is under increasing pressure to address unsustainable practice. Emerging out of the pandemic era the fashion industry is also responding to a variety of complex industry challenges such as high return rates, customer demand for better-fitting apparel, faster fashion, the drive towards personalisation and greater transparency and sustainability across the value chain. These factors along with increasing labour costs are furthermore exerting force on the industry to embrace nearshoring and reshoring. Based on extensive primary research involving oral histories methodology with leading industry professionals involved in the innovation of technological and digital solutions for the fashion industry, this book presents the latest advances in sewn product technology which offer solutions to many of the fashion industry's current and emerging challenges whilst also informing how these developments are influencing fashion jobs of today and tomorrow. This book is therefore of value to fashion students, academics, researchers, and technicians as well as those working within the fashion industry involved in the design, development, manufacture, buying and retail of fashion apparel.

Features:

- Provides a comprehensive insight into the latest advances in sewing machine technology including advanced automation and robotics used in the manufacture of fashion apparel.
- Provides a comprehensive insight into the latest industrial sewing threads and needles that can effectively support sustainable design practice.
- Exclusively covers advances in digital technologies to support sustainable practice including advances in 3D body scanning and digital measuring systems, recent advances in digital pattern making and pattern design systems, recent advances in 3D fashion design software and the latest advances in Product Lifecycle Management (PLM) systems used within the fashion industry.
- Includes sections on advances in no-sew seam bonding and ultrasonic welding technologies.
- Provides an insight into advancements in 3D cloth simulation and prototyping for apparel design and gaming.
- Enables readers to understand the impact of the latest advances in sewn product technology on the jobs of today and tomorrow.
- Case studies that provide working examples of advances in sewn product technology.

Advances in Sewn Product Technology

Anita Mitchell

CRC Press
Taylor & Francis Group
Boca Raton London New York

CRC Press is an imprint of the
Taylor & Francis Group, an **informa** business

Designed cover image: © Style3D

First edition published 2024
by CRC Press
2385 NW Executive Center Drive, Suite 320, Boca Raton FL 33431

and by CRC Press
4 Park Square, Milton Park, Abingdon, Oxon, OX14 4RN

CRC Press is an imprint of Taylor & Francis Group, LLC

ISBN: 978-0-367-64825-1 (hbk)
ISBN: 978-0-367-64828-2 (pbk)
ISBN: 978-1-003-12645-4 (ebk)

DOI: 10.1201/9781003126454

Typeset in Times
by Apex CoVantage, LLC

This book is dedicated to my beautiful little Charlotte

Contents

About the Author

Anita Mitchell is Principal Lecturer, division head fashion design and head of digital fashion technologies at Manchester Fashion Institute at Manchester Metropolitan University, one of the largest fashion schools in the UK. She is Senior Fellow of the Higher Education Academy and holds external examiner and principal external examining posts at other key fashion institutes in the UK. Having worked in higher education for nearly two decades lecturing in the areas of fashion product development, fashion technology, sewing machine technology, digital fashion technology and automation, her interest in pedagogy led her to the role of Associate Dean for Teaching and Learning whilst leading a variety of key projects such as the development of the fashion innovation lab within the Manchester Fashion Institute and a variety of higher education and curriculum-related reviews. She is currently completing doctoral research towards an EdD whilst leading a high-profile digital fashion strategy that is transforming higher education fashion curriculum and future proofing the graduate skills needed to support the ongoing digitalisation of the fashion industry. In terms of her writing and publication experience, Anita has contributed to a variety of fashion technology books and authored book chapters in areas that include fashion product development, fashion technology and advanced joining technology.

Anita began her career working as a designer for a well-known menswear brand in the UK and moved on as product developer for a leading British knitwear specialist whilst studying at the London College of Fashion. This facilitated her career move to senior fit technologist and technical manager for one of the largest suppliers to the British clothing industry. Her product specialism spans light sewing, ladies and men's formal wear, casual wear, sports, and leisurewear. She has been involved in developing a variety of highly technical apparel products such as pre- and post-cure non-iron apparel and wet processing techniques in some of the world's most highly automated factories. She was also involved in trailblazing technologies such as seam-free bonding and welding technology and the use of 3D digital fit technology to support agile and sustainable product generation. She has developed apparel for brands such as Ralph Lauren, Ted Baker, and Hussein Chaylan. Her experience of offshore manufacturing afforded her the opportunity to work as technical manager in many locations around the world, such as Portugal, Morocco, and Malaysia. Anita has also been fortunate to work alongside British designers such as Katherine Hamnett when she supported the development of one of the first organic collections for Marks and Spencer Autograph range. This role also afforded her the opportunity to develop novel and innovative trims with some of the world's leading trimmings suppliers.

Preface

In order to fully contextualise both the current relevance of this book and also the rationale and motivation for writing it, I thought it important to begin by briefly explaining why the varieties of new and emerging advances in sewn product technology are critical in supporting the fashion industry to restructure through the current period of immense disruption and change brought about by the Fourth Industrial Revolution.

The Fourth Industrial Revolution, also known as 4IR, began in Germany at the beginning of the millennium triggered by Germany's national strategy of 2009, called the Digital Agenda. The Digital Agenda was a national master plan, involving industry, academia, and the government, designed to overcome the economic crisis that was happening in Germany. During this period, a minority of forward-thinking enterprises had already begun to embrace the move towards advanced digitalisation, advanced connectivity, and advanced automation to support sustainable productivity. The term Industry 4.0 was used to describe this new industrial change in the title of a report presented to the German government in 2009.

In terms of progress, 4IR is quite unlike all preceding industrial revolutions as the technologies associated with this revolution are subject to three distinct waves of disruption forecasted for implementation within an incredibly short 16-year time frame from 2009 to 2025. The first wave was initially brought about by the Third Industrial Revolution involving digitalisation, advanced analytics, cloud computing, augmented reality, robotics, and 3D printing. However, the use of many of these technologies has only recently begun to be adopted by the fashion industry accelerated by pressing global issues, including the recent global pandemic, which has necessitated many of the fashion industry's customary processes, involving design, product development, and manufacturing, to rapidly change in a record-breaking time span. Astoundingly, during the recent global pandemic, some fashion industry leaders declared two years of digital transformation in two months. In this short time frame, digital technology has proven to offer a variety of enhancements and benefits to the industry whilst also spawning a variety of other new developments in the areas of advanced automation and robotics. For example, the use of cloud computing with SaaS-based software has increased dramatically as the wider benefits of 3D prototyping software were realised during the pandemic when physical sampling was not possible.

It is unquestionable that the fashion industry is behind in the adoption of many of these technologies. Having passed into 2024, some of the wave two and three technologies needed to drive 4IR such as 6G communications are just on the horizon. Smart Automation are predominantly in pilot phase, AI is accelerating much faster whilst blockchain is gaining momentum and becoming increasingly important for tracking the fashion supply chain and delivering transparent data to the consumer concerning the social compliance and sustainability of apparel supply chains.

Emerging out of the pandemic era, fashion brands are also responding to a variety of complex industry challenges such as reducing the high return rates within the apparel e-commerce sector with opportunities to use digital body scanning technologies, big data, and AI to develop apparel that better fits the consumer or for fulfilling smaller more complex orders driven via near real-time consumer demand. The new consumer not only expects faster fashion with smaller, more personalised collections with options for customised or made-to-measure apparel but also expects greater transparency and sustainability across the value chain. Paradoxically, in response to sustainability agenda, this is also resulting in a move towards non-seasonal slow fashion and transparent circular supply chain practices that can honestly meet due diligence and social compliance. These factors along with increasing labour costs are furthermore exerting pressure on fashion businesses and manufacturers to locate apparel manufacture closer to retail by moving away from offshoring to nearshoring and reshoring.

Based on extensive primary research involving oral histories methodology with leading industry professionals involved in the innovation of technological and digital solutions for the fashion industry, this book presents the latest advances in sewn product technology which offer solutions to many of the fashion industries current and emerging challenges whilst also informing how these developments are influencing fashion jobs of today and tomorrow. This book is therefore of value to fashion students, academics, researchers, and technicians as well as those working within the fashion industry involved in the design, development, manufacture, buying, and retail of fashion apparel.

Acknowledgements

The author would like to express her deepest gratitude to all contributors listed below for their valuable time, experience, and knowledge.

CHAPTER 1   ADVANCES IN SEWING MACHINE TECHNOLOGY

Key sewing machine exhibitors at the Texprocess 2022 exhibition, Frankfurt, including PFAFF, Vi BE Mac, Juki, Strobel, Durkopp Adler, Efatech (Turkey), and Vetron.

CHAPTER 2   ADVANCES IN STITCHING NEEDLE TECHNOLOGY

Key contributors include Alexander Moser: Senior BD Manager of Industrial Sewing at Schmetz Needles, and Jochen Gunther Menger: Senior Manager, Product Development Sewing and Tufting at Groz-Beckert.

CHAPTER 3   ADVANCES IN SEWING THREAD TECHNOLOGY

Key contributors include Coats Group PLC, who are the world's largest manufacturer and distributor of sewing thread, and A&E Gutermann Threads, and Twine, a leader in on-demand thread dyeing sytems.

CHAPTER 4   ADVANCES IN NO-SEW SEAM BONDING AND ULTRASONIC WELDING TECHNOLOGIES

Key contributors include Tony Turner and Nick Turner (Sew Systems), Dave O'Leary, Wullie Daly (Ardmel Automation), Bill Reece (Macpi), and Dr Steven G. Hayes (University of Manchester).

CHAPTER 5   AUTOMATION AND ROBOTICS IN THE APPAREL MANUFACTURING INDUSTRY

Key contributors include Jonas Hillenburg Vention and representatives from PFAFF, JUKI, Durkoff Adler, Yamato, Mitsubishi, and Vi BE Mac, who were interviewed at the Texprocess industrial machinery exhibition in Frankfurt in June 2022.

CHAPTER 6   RECENT ADVANCES IN 3D BODY SCANNING AND DIGITAL MEASUREMENT SYSTEMS

Key contributors include Richard Barnes (CEO and Founder of Select Research), Richard Allen (Shape Analysis), Alex Chung and Matthew McMillion (Artec 3D) along with leading experts and CEOs from three of the world's leading body scanning developers, including Dr Mike Fralix (TC2), Dr Helga Gaebel (Avalution: Human Solutions/Humanetics), and David Bruner (Size Stream).

CHAPTER 7   RECENT ADVANCES IN DIGITAL PATTERN MAKING SYSTEMS

Key contributors include Mathieu Bonnenfant, Vice-President of Product Marketing, Lectra.

CHAPTER 8   RECENT ADVANCES IN 3D FASHION DESIGN TECHNOLOGY

Key contributors include Rebecca Ma (Digital Marketing Director) Style3D, Eric Liu (Founder and CEO) Style3D, Alfie Chen (Chief Operating Officer and Chief marketing Officer) Style3D, and Elizabeth Brandwood, Style3D, Zhejiang Linctex, Digital Technology, China.

CHAPTER 9   THE GAP BETWEEN DIGITAL AND PHYSICAL 3D PROTOTYPING FOR FASHION

Key contributor is Kate Ryabchykova, 3D Fashion Prototyping Artist and Lecturer at Manchester Fashion Institute at Manchester Metropolitan University. Other minor contributors include Professor Mykola Riabchykov, specialist in sewing equipment and technologies at the Lutsk National Technical University, Ukraine; Vernita Sothirajah, a commercial fashion designer; and Martina Harvey Capdevila, the owner of the sustainable brand Original Source and Supply.

CHAPTER 10   ADVANCES IN 3D CLOTH SIMULATION AND PROTOTYPING FOR APPAREL DESIGN IN GAMING

Key contributors include Adam Cain (Lecturer of Games Art, School of Digital Arts at Manchester Metropolitan University), also minor contributors, including Harvey Parker (Art Director at the Multiplayer Guys) and Shiloh Ragins (Lecturer of Games Art at Confetti Institute of Creative Technologies, at Nottingham Trent University).

CHAPTER 11   ADVANCES IN PRODUCT LIFECYCLE MANAGEMENT (PLM) SYSTEMS WITHIN THE FASHION INDUSTRY

The key contributor for this chapter, Mark Harrop, often dubbed the PLM guru, is a highly respected fashion industry PLM expert and CEO of a leading industry-focused PLM market analysis publication called *WhichPLM*.

CHAPTER 13   VORTEQ CASE STUDY: HIGH-PERFORMANCE CYCLEWEAR PRODUCT DEVELOPMENT

Key contributors to this chapter include Dr Rob Lewis (OBE), founder of Vorteq and Managing Director of TotalSIM, leaders in aerodynamics and computational fluid dynamics, and Jane Ledbury, an emeritus apparel industry expert and principal lecturer in fashion design and product development at Manchester Metropolitan University.

CHAPTER 14 FOCUS BRANDS CASE STUDY: IMPLEMENTING A PLM SYSTEM

Key contributors include Focus Brands employees Timothy Peck (Head of Design), Rebecca Minors (Design Team Manager), and investigating for this case study Jennifer Collinson (Production and Sourcing Manager).

1 Advances in Sewing Machine Technology

1.1 INTRODUCTION

This chapter is concerned with advances in industrial sewing machines that are not only considered the cornerstone of the fashion industry's manufacturing sector, but they are also the essential building blocks from which semi-automated, automated, and advanced sewing systems are developed and built from. This chapter will outline recent advances in sewing machine technology and how these developments are being driven by a variety of key disruptors, including rising labour costs, consumer demand for personalised apparel, the digitalisation of the fashion industry, sustainable manufacturing, as well as social compliance and worker welfare. The chapter will close by considering how sewing machine advancements will assist in shaping the future of apparel manufacturing. To provide the most current insights regarding recent and emerging advancements in this area, key machine makers such as PFAFF, Vi BE Mac, Juki, Strobel, Durkopp Adler, Efatech (Turkey), and Vetron were interviewed during the 2022 International Texprocess sewing machinery exhibition in Frankfurt.

1.2 KEY DISRUPTORS DRIVING SEWING MACHINE ADVANCEMENTS

Since its introduction in the 1880s, the electrical–mechanical sewing machine, utilising gears and cams, remained relatively unchanged for nearly 100 years and was regarded as the single most important tool in the fashion manufacturing industry until the advent of circuit boards and microchips in the 1980s which enabled the development of programmable microprocessor-controlled sewing functions such as needle positioning, thread trimming, or programmable stitch count. These advancements were considered groundbreaking at the time in terms of their ability to enhance quality- and cost-based efficiencies whilst also supporting advancements in sewing machinery automation from the 1980s onwards which also assisted to temporarily support cost-effective apparel manufacturing in the UK but only until the mid-1980s, when many manufacturers offshored garment making to low labour cost countries in the East, and in particular Asia, to counteract the impact of cheaper import penetration (Jones, 2006).

Since the late 1990s and early millennium period, both the rapid advancements and cost reductions in computing hardware and software have greatly supported a plethora of advancements in digital sewing machine technology to the point where every function of a sewing machine such as speed, stitches, thread tension, feed,

DOI: 10.1201/9781003126454-1

presser foot tension, and diagnostics for quality and maintenance can be computer controlled (Jana, 2020).

Despite these revolutionary developments, the abundance of cheap offshore labour has until recently minimised pressure on manufacturers, particularly small and medium-sized enterprises (SMEs), to invest in sewing machine advancements to upgrade garment manufacturing. However, this situation is changing due to a number of interrelated disruptors, the most pressing of which includes increasing labour costs which continue to rise in the lowest cost regions of the world. This is threatening the future competitiveness of key manufacturing locations such as Asia as nearshoring and reshoring pose tangible solutions to both cost and sustainable agenda.

The post-pandemic fashion industry and also the upstream functions of buying and merchandising are becoming increasingly digitalised, disrupting traditional supply chains whilst also resulting in the development of new models of apparel production such as micromanufacturing, as well as the movement away from offshoring to nearshoring and onshoring. This is enabling fashion brands to buy and produce smaller, more complex orders that are being driven by near real-time consumer demand (Harrop, 2022).

Concurrent sewing machine advancements are also needed to meet the manufacturing challenges posed by the demands of the new consumer, regarded as a disruptor for their expectations of faster personalised collections more frequently; they also expect greater transparency and sustainability across the value chain. This is also paradoxically resulting in a movement towards non-seasonal slow fashion and circular supply chain models. The consumer is also directly driving sustainability and social compliance which compels brands to disclose their environmental and social impact and track worker welfare all the way through the value chain to now include apparel manufacture, which is an area that has traditionally remained hidden from the consumer.

These disruptors and the challenges they present to the apparel manufacturing sector are directly informing new developments in sewing machine technologies, and throughout the rest of this chapter, each of the key disruptors outlined in the following bulleted list will be further considered and illustrated using examples of recent sewing machine advancements.

- Rising labour costs
- Demand for personalised apparel and the development of new manufacturing models such as micromanufacturing
- Digitalisation of the fashion industry, supporting sustainable manufacturing, nearshoring and onshoring
- Social compliance and worker welfare

1.3   SEWING MACHINE ADVANCES IN RESPONSE TO RISING LABOUR COSTS

For over 50 years, Asia has remained the main garment manufacturing centre of the world due to the availability of low-cost labour. Countries such as China, Malaysia, and Thailand have prospered during this period and experienced rapid economic growth and a movement away from manufacturing. However, labour costs in countries located in East and Southeast Asia, such as Bangladesh, Cambodia, Vietnam, and Myanmar, have been significantly lower but are now rising when compared to developing and emerging economies such as Bulgaria, Romania, Hungary, Poland, and Turkey. In 2019, East and Southeast Asian garment manufacturers employed 60 million workers and accounted for 55% of global textiles and clothing exports (ILO, 2022). Most garment manufacturers in East and Southeast Asia are involved in cut, make and trim (CMT), and most are reported to have low levels of productivity using low-skilled labour and low-tech machinery and equipment, which when combined with falling garment prices and rising minimum labour costs is resulting in narrow profit margins of around 5% making it unviable for SMEs to invest in restructuring or upgrading technology, which could assist improving productivity, quality, and profit margins (ILO, 2022). According to Lee (2016), 'even limited increases in wages can make thin profit margins disappear' (ILO, 2022, p. 22). The average minimum wage in East and Southeast Asia is rising due to both mandatory minimum wage legislation and growing consumer demand for proof of environmental and social improvements in value chains.

Governments in countries such as Cambodia, Vietnam, and the Philippines are also under pressure from unions to increase wages as they seek to prevent outbreaks of social unrest that could cause political instability. In 2015, the average minimum wage in East and Southeast Asia was just 63% of the global average. In less than five years, minimum wages have risen to 82% and are due to increase further. For example, Laos, China, and Vietnam had the largest year-on-year average minimum wage growth rates of 14.6%, 9.8%, and 8.8% respectively between 2015 and 2019. In Vietnam, the 2019 minimum wage also increased by an average of 5.3% across its four regions (Fitch Solutions, 2019). Projections indicate that the minimum wage in most of Asia's garment manufacturing countries may overtake global average minimum wage levels within the next ten years and therefore threaten Asia's manufacturing competitiveness. Rising labour costs and reduced lead times are forcing fashion brands to re-evaluate their supply chains and shift manufacturing closer to the point of retail through strategies that involve nearshoring and onshoring to countries such as Bulgaria, Romania, Hungary, Poland, and Turkey (ILO, 2022). It is therefore important for Asian clothing manufacturers to invest in new technologies that can improve efficiencies by stepping up productivity, reducing human error, and improving quality, waste, and speed to market (ILO, 2022; Fitch Solutions, 2019). There are already successful examples in Asia to demonstrate this approach works. For example, China has already begun the journey to reduce reliance on labour and is aggressively investing in automated sewing machine technology to maintain in-country operations that are further enhanced by the country's integrated vertical supply chain (Lee, 2016).

There are several recent sewing machine advances that assist to reduce labour costs. For example, Juki has introduced the AMS-224EN6030/AW-3/shuttle bobbin machine that supports the automatic winding of a spare bobbin and the automatic changing of the bobbin whilst the machine is still sewing. The machine can also remove and dispose of any remaining thread left on the empty bobbin. This means that there can be continual sewing and no downtime associated with winding or changing bobbins.

FIGURE 1.1   Juki AMS-224EN6030/AW-3/shuttle bobbin.

Source: Courtesy of https://jukieurope.com

An example demonstrating how advancements in sewing machinery can reduce labour costs associated with reworking defective seaming is the PFAFF 5483–911/35 sewing machine. This machine has been designed for joining the back rise on trousers using a 401 double lock chainstitch, with the advancement of a back tack function. This is an unusual advancement as most chainstitches, which can unravel easily, do not have this back tack function, and so this innovation helps in eliminating this quality risk and indeed the work associated with reworks.

1.4   SEWING MACHINE ADVANCES IN RESPONSE TO DEMAND FOR PERSONALISED APPAREL AND THE MOVE TOWARDS NEW MODELS OF MANUFACTURE

The origins of personalisation within the fashion industry pre-date the advent of mass production, at a time when most items of clothing were made for individuals by tailors or seamstresses or were self-made by individuals who could not afford to pay for make services. The growth of mass manufacture at the end of the Second World War led to the mass market of products and mass consumerism. Personalisation during this time signified expensive exclusivity that was often reserved for the wealthy; however, the introduction of digital technologies in the mid-1990s and early millennium enabled the fashion industry to make available personalised apparel to the masses. Examples include Levi's Original Spin Programme using 3D body scanners to support the creation of custom-fit jeans. More recently, the availability and accessibility of digital technologies have continued to accelerate as their cost continues to decrease enabling fashion brands from the high street to designer and luxury sectors to meet consumer demand for personalisation.

Digital technologies are now able to not only influence consumer purchasing but also predict what consumers will buy before they make a purchase. Moreover, they can be used to efficiently and cost-effectively design, develop, manufacture, and deliver bespoke products to fashion consumers who have increasingly high expectations for personalisation.

Despite the fact that the fashion industry has recently entered a period of instability due to factors such as the war in Ukraine, rising interest rates, and inflated energy prices which are likely to impact expenditure on clothing in 2023 (McKinsey, 2022), demand for personalised apparel particularly amongst middle- and higher-income earners is likely to continue, and growth of this market is evident with prominent retail chains such as the UK-based Flannels establishing a made-to-measure tailoring hub devoted to made-to-order tailoring (BOF, 2022). Other examples include Levi's Lot 101, which offers a bespoke service to create the perfect hand-tailored jeans, or Ralph Lauren's custom service which enables customisation of the classic polo shirt or cap with print or embroidery. According to Kersnar (2022), opportunities for luxury brands to develop new fashion markets that can offer personalised and dedicated collections have also been realised in locations such as the Middle East by extending personalisation strategies to meet the specific requirements of localised consumers.

The move from mass production to mass customisation requires new manufacturing models that can quickly, efficiently, and cost-effectively produce both small and individual orders using the required combinations of digital technologies and advanced automation that are discussed in more detailed in a later chapter. However, there are also a number of new sewing machine advances that have been designed to support new manufacturing models for the production of both small and individual orders such as the Jack A10+ Lockstitch Sewing Machine. This is a highly flexible and intelligent integrated sewing system that can be linked to a factory's production management system to communicate real-time data relating to production progress as well as intelligent remote-service-requests, for example there is a fault with the machine. The sewing machine can also connect with a factory's hanging system to acquire detailed information about the next product that will be presented for construction such as fabric characteristics. A range of optimum sewing machine settings can also be pre-programmed or recalled. Other features include four types of feed dogs that can be changed digitally, and unlike traditional lockstitch sewing machines, it can be digitally set to sew light, medium, and heavyweight materials without the need for time-consuming mechanical adjustments. The design and flexibility of this machine are important developments for factories involved in the manufacture of made-to-measure apparel where an order can be as small as one single bespoke garment.

Juki has also produced a product similar to the Jack A10+ known as the Juki DDL-9000C—a digitalised lockstitch machine which has been designed for the made-to-measure manufacturing market. Unlike traditional lockstitch machines which are adjusted by operators or sewing machine mechanics based on personal knowledge and experience, the DDL-9000C sewing machine enables fully digitalised adjustments to be made remotely either to a single machine or to a bank of

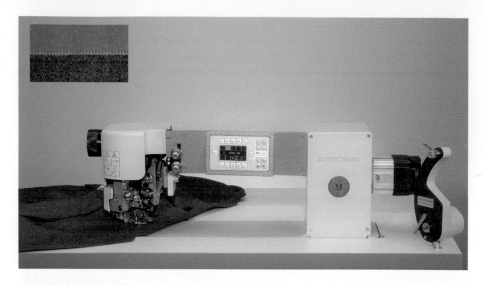

FIGURE 1.2   Strobel 718.

Source: Courtesy of Strobel www.Strobel.com

machines from a smart tablet, meaning settings for each individual fabric can be stored and reproduced with total accuracy each time the same material is utilised. This machine has both a vertical- and a horizontal-driven digital feed mechanism, meaning the height of the feed dog and the feed locus can be digitally adjusted to suit different weights of fabric, and both the needle tension and the presser foot pressure are also digitally controlled.

With the rise in demand for bespoke made-to-measure apparel, including bespoke tailoring, there have been numerous developments in the area of simulated hand-stitch technology such as the Strobel 718, which delivers two different stitches at the same time. The first row is a lockstitch, and the parallel row is a chainstitch which simulates the hand-sewn T-stitch traditionally used in the construction of the Melton undercollar.

Likewise, the AMF Reece DECO 2000 sewing machine can simulate classic hand-tailored stitching to include long/short variations of pick stitch and saddle stitch along with a customised selection of decorative stitching.

Flexible machinery is a key requirement for manufacturing small or individualised orders, particularly if a machine can be quickly and efficiently converted to produce a different stitch, as this can reduce overhead costs by minimising the need to purchase more machines that will take up more space. ViBeMac has developed a highly innovative solution with its Vii.BeE.MAC—Model 3022 BHE. This machine has been around for many years, but its relevance has been underplayed until now. The machine can be converted from a lockstitch to a chainstitch and takes a trained mechanic just 12 minutes to convert. This machine is specifically relevant for micro-manufacturing due to its flexibility.

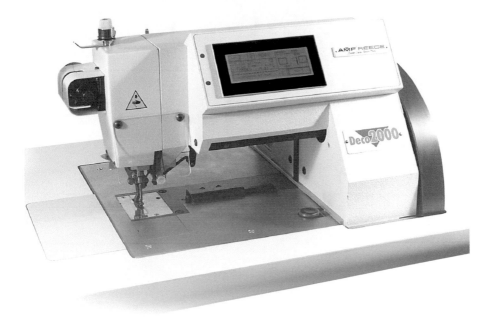

FIGURE 1.3   AMF Reece DECO 2000 sewing machine.

Source: Courtesy of AMF Reece www.amfreece.cz

1.5   SEWING MACHINE ADVANCES

The digitalisation of the fashion industry has been characterised by a fusion of technologies that include, for example, high-speed internet, AI, advanced automation and
robotics, big data analytics, and cloud technology. The significant change brought
about by these technologies has been revolutionary, with benefits that include the
transformation of the supply chain via nearshoring and onshoring through technological innovation with gains in efficiency and productivity, leading to a reduction in
transportation and communication costs, as well as the opening up of new markets
and economic growth (Schwab, 2015).

A recent advancement that can assist with efficiency gains in apparel sewing productivity is a production monitoring solution for sewing factories called
QONDAC. Produced by Durkopp Adler, a leading sewing machine developer,
QONDAC can link up to 1,500 sewing machines in a network and analyse the
productivity and status of each machine with access to real-time data, assisting
to determine if productivity is in line with critical paths. The software can also
provide data on machine performance via SAP or PLM platforms, communicating
downtime and fault detection via automatic alarm messages that can be sent digitally to smartphones of supervisors, mechanics, or managers. Individual machines
can be adjusted centrally from the server, and maintenance processes can be digitised based on real usage (OCS, 2018).

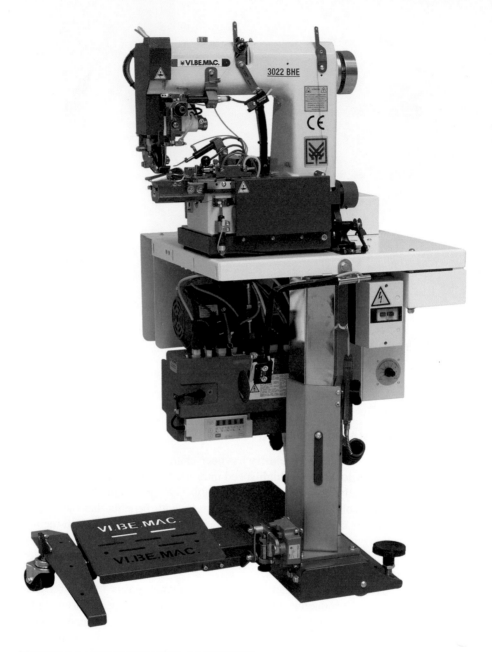

FIGURE 1.4   Viii.BeE.MAC Model 3022 BHE.

Source: Courtesy of www.vibemac.com.

Similar to QONDAC, Turkish-based EfaTech has launched TRACETECH aimed at manufacturers who are involved in nearshoring and reshoring. The benefit of the system includes the reduction in labour as less supervisors are needed to monitor and track performance in large sewing factories and up-to-date data is available to inform overall progress with orders and completion rates.

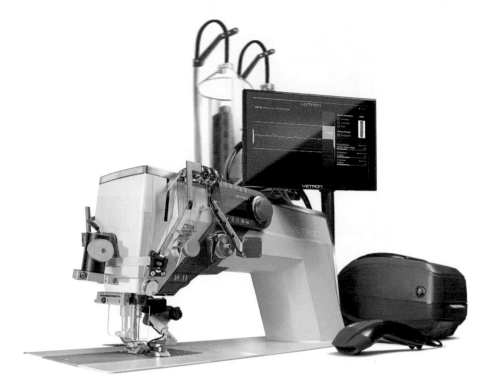

FIGURE 1.5   Vetron 5700 digital sewing machine.

Source: Courtesy of Vetron www.vetrontypical-europe.com

Advancements in sustainable sewing machine design and flexibility have become increasingly popular specifically as the industry responds to changing models of manufacture involving smaller orders with continual style change where downtime associated with machine maintenance must be minimised. Vetron was one of the first sewing machine manufacturers to launch a fully modular sewing machine range whereby all machine components are grouped into 'technology modules' where module/parts can be exchanged at any time making it possible, for example, to convert a flatbed machine into a longarm machine. This advancement won the Texprocess innovation award over a decade ago and other leading machine makers such as Durkopp Adler have since developed modular sewing machine ranges such as the M-Type (Vetron, 2017; Jana, 2020).

1.6   SEWING MACHINE ADVANCES IN RESPONSE TO SOCIAL COMPLIANCE AND OPERATOR WELFARE

Global apparel manufacturing is generally hidden from the public eye, and there are many cases of unsafe working conditions accompanied by low wages where workers generally remain powerless because trade unions are often suppressed. Consumers have become increasingly aware and concerned about this situation, and they along with fair labour organisations have been lobbying brands and manufacturers to implement transparency measures in making information regarding working conditions in

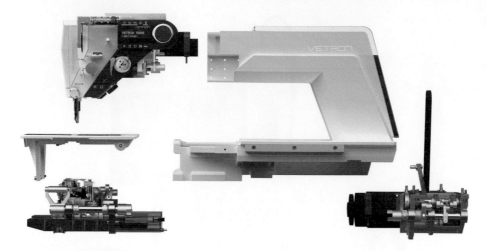

FIGURE 1.6 Vetron flatbed explosion.

Source: Courtesy of Vetron www.vetrontypical-europe.com

sewing factories publicly available, and this is assisting to improve the conditions that workers experience in the global apparel manufacturing industry (LBL, 2022).

The demand for transparency and support of worker welfare is having a positive effect with more sustainability and social compliance regulations coming into force across many nations. Laws governing due diligence have already been passed in some countries mandating that companies have visibility over their supply chains, and there are also several mandatory regulations that have already been passed by governments in the United States, Germany, France, and Norway, such as the US Uyghur Forced Labour Prevention Act, New York Fashion Sustainability and Social Accountability Act, German Supply Chain Due Diligence Act (LkSG), French Duty of Care Law, and Norwegian Transparency Act.

More recently, many UK companies, investors, and business associations, which include John Lewis, Tesco ASOS, Primark, Unilever, The British Retail Consortium but to name a few, have called on the UK government to introduce a law that mandates companies to carry out human rights and environmental due diligence.

There are many examples of brands and manufacturers that strive to enhance the welfare of sewing machinists by designing features that make sewing more comfortable, protecting the user from fatigue or injury. For example, Vi.BeE.Mac has introduced a sewing machine pedal design that has an integrated shock absorption system to reduce operator fatigue and injury caused by vibrations that can result in long-term damage to bones and joints. Vi.BeE.Mac has also switched to manufacturing their sewing machines and workstations to have height-adjustable tables so the machine can be altered as they have realised people in different countries are of different heights, and this simple benefit can support more comfortable working conditions and better productivity.

A further example of a sewing machine advancement designed to support workers' welfare is the development of pedal-less sewing which can enhance operator comfort and reduce fatigue. Instead of a pedal, sensors are built into the sewing machine which sense-track the operator's movement enabling the automatic engagement of the sewing

FIGURE 1.7   Vetron Flatbed Explosion.

Source: Courtesy of Vetron www.vetrontypical-europe.com

FIGURE 1.8   Vetron VE5656 Autoseam.

Source: Courtesy of Vetron www.vetrontypical-europe.com

function. Vetron, a leading sewing machine developer, has introduced the *Vetron Trace*, which is a single-needle lockstitch fitted with specialist sensors that enable automatic sewing based on hand-movement mapping. This advancement won a Texprocess Innovation Award in 2017 for its ability to capture the natural working movements of the users to support automatic sewing (Apparel Resource, 2018; TexProcess, 2017).

A similar development is the Jack C4 overlock machine, which offers three sewing modes including a fully automatic mode similar to the Vetron 5656 auto seam machine that can be used, for example, to join long straight seams. Just like the Vetron Trace model, the inbuilt sensors map hand movement which initiates the sewing function cameras located in the bottom plate and machine head. A special top and bottom feed automatically feeds materials without the need for a human operator.

1.7   SEWING MACHINE ADVANCES FOR THE FUTURE

Since its introduction in the 1800s, the sewing machine utilising needles and thread, supported by ongoing advancements involving mechatronics and more recently digital technologies, has remained the most efficient method of joining material seams despite the introduction of other more specialist joining methods such as seam bonding and ultrasonic welding.

Beyond the elite smart factories of the world synonymous with names like Levi's, Mass Holdings, or Hugo Boss, Asia remains for now, the garment-manufacturing centre of the world, composed predominantly of SMEs using low-tech sewing machines and equipment. The standard lockstitch, overlocker, and cover seam therefore remain the backbone of the global apparel manufacturing industry and the essential building blocks from which all sewing machine advancements including semi-automated, automated, and advanced sewing systems are developed and built from.

As the global apparel industry continues to transform, challenges such as increases in labour costs and the movement towards manufacturing closer to market will continue to emphasise the need to invest in sewing machine advancements (including advanced automation that will be discussed in a later chapter) that can progressively eliminate human labour from garment construction whilst meeting the diverse and increasingly personalised requirements of the new consumer.

The examples given earlier in this chapter demonstrate the shift towards operatorless apparel manufacturing supported by digitalisation of sewing machinery such as the introduction of programmable and vision-based sewing machinery used for decorative stitching, simulated hand stitching that mimics bespoke tailoring skills, and pedalless machines where stitching is controlled electronically rather than by the operator. In addition, automatic bobbin changers, machine-to-machine communication, digital monitoring systems that support predictive maintenance along with voice-guided features that can inform the nature of faults are all forecasted to become the norm in sewing factories, making the digital factory a more widespread reality (Jana, 2020).

Ongoing growth within the global sewing machine market indicates that investment is taking hold as this market has not only remained buoyant, but it is also forecasted to grow further. Its size was estimated at US\$4,278.35 million in 2021 and is projected to grow at a compound annual growth rate (CAGR) of 7.20% to reach US\$6,494.79 million by 2027 (Business Wire, 2022).

Successful apparel manufacturers of the future will be those who aggressively invest in new digital sewing machine technologies that, when combined with advanced automation, robotics, and other advantages such as operating within a vertical supply chain, can reduce the overall reliance on human labour, increase productivity, reduce human error, and improve quality, waste, and speed to market.

REFERENCES

Apparel Resource (2018) *10 Popular Sewing Technology Trends* [online], https://apparelre sources.com/technology-news/manufacturing-tech/10-popular-sewing-technology-trends/ (Accessed 14/12/2022)

BOF (2022) *The Unlikely Return of Bespoke Suiting, Explained* [online], www.businessof fashion.com/articles/retail/the-unlikely-return-of-bespoke-suiting-explained/ (Accessed 16/12/2022)

Business Wire (2022) *The Worldwide Sewing Machine Market Research Report* [online], www.businesswire.com/news/home/20220713005773/en (Accessed 29/12/2023)

Fitch Solutions (2019) *Rising Labour Costs Gradually Eroding East and South East Asia's Competitiveness* [online], www.fitchsolutions.com/bmi/country-risk/rising-labour-costs-gradually-eroding-east-and-south-east-asias-competitiveness-03-05-2019#:~: text=From%20a%20labour%20costs%20perspective,increasingly%20higher%20costs %20of%20production (Accessed 19/12/2022)

Harrop, M (2022) PLM Report 2022. *The Interline* [online], https://theinterline.com/PLM-Report-2022.pdf (Accessed 11/11/2022)

ILO (2022) *Employment, Wages and Productivity Trends in the Asian Garment Sector: Data and Policy Insights for the Future of Work* [online], www.ilo.org/asia/publications/ WCMS_848624/lang--en/index.htm (Accessed 02/01/2024)

Jana, P and Jung, H (2020) *Technology Evolution in Apparel Manufacturing, Apparel Resource*, Kindle E-Book [online], https://books.google.co.uk/books/about/Technology_Evolution_ in_Apparel_Manufact.html?id=TZbqDwAAQBAJ&redir_esc=y (Accessed 02/01/2024)

Jones, RM (2006) *The Apparel Industry*, 2nd ed., Blackwell, Oxford

Kersnar (2022) *The State of Fashion 2023: Holding onto Growth as Global Clouds Gather* [online], www.studocu.com/in/document/institute-of-engineering-and-management/ computer-science-engineering/the-state-of-fashion-2023-holding-onto-growth-as-global-clouds-gathers-vf/69062878 (Accessed 02/01/2024)

LBL (2022) *Transparency: What You Know You Can Change* [online], https://labourbehindthe label.org/transparency/ (Accessed 02/01/2024)

Lee, SS (2016) *Creating Decent Work and Sustainable Enterprises in the Garment Sector in an Era of Change: Employers Perspectives* [online], https://www.ilo.org/wcmsp5/ groups/public/---asia/---ro-bangkok/documents/meetingdocument/wcms_579471.pdf (Accessed 02/01/2024)

McKinsey (2022) *Economic Conditions Outlook During Turbulent Times, December 2022* [online], www.mckinsey.com/~/media/mckinsey/business functions/strategy and corpo rate finance/our insights/economic conditions outlook 2022/december 2022/economic-conditions-outlook-during-turbulent-times-december-2022.pdf (Accessed 11/12/2023)

OCS (2018) *Qondac Networks, Production Monitoring Solutions by Durkopp Adler* [online], www.onlineclothingstudy.com/2018/10/qondac-networks-production-monitoring.html (Accessed 22/12/2022)

Schwab, K (2015) *The Fourth Industrial Revolution: What It Means and How to Respond* [online], www.weforum.org/agenda/2016/01/the-fourth-industrial-revolution-what-it-means-and-how-to-respond/ (Accessed 02/01/2024)

Texprocess (2017) *Innovation Award* [online], https://home.nestor.minsk.by/exhibitions/ news/2017/05/1101.html (Accessed 02/01/2024)

Vetron (2017) *Reflect Technology Modular System* [online], www.reflect.com.tw/?page_id= 2327 (Accessed 02/01/2024)

2 Advances in Stitching Needle Technology

2.1 INTRODUCTION

This chapter will provide a brief overview of the development of sewing needles, followed by a review of needle design for diverse stitching technologies. The chapter will move on to explore the current key drivers of innovative needle design and development, including the drive towards sustainable design practice, sustainable sewing needle supply chains, faster fashion, developments in new textiles, such as technical textiles and trims, micromanufacturing for personalised and made-to-measure apparel, and sewing automation. The chapter will close with a detailed review of recent advances/innovations in stitching needle design and how these are responding to current sewing technology challenges in the apparel industry.

To provide the most current insight regarding recent and emerging advancements, key contributors include Alexander Moser: Senior BD Manager of Industrial Sewing at Schmetz Needles, and Jochen Gunther Menger: Senior Manager, Product Development Sewing and Tufting at Groz-Beckert. Both Schmetz and Groz-Beckert are world leaders in sewing needle design, development, and manufacture.

2.2 BACKGROUND OF THE SEWING NEEDLE

The sewing needle was one of the first tools used by man, and over the centuries it developed from a simple craft item to a precision tool for modern sewing machines. It has been constantly adapted for new industrial applications from fashion to performance garments, furnishings, car seats, and in the construction of products that require high technical safety standards, such as airbags, anti-G-suits, and even astronaut suits. Most ancient sewing needles, which date back to 28000 BC, did not have an eye but a split end which gripped the thread to be sewn (often raffia, gut, or sinew). Needles from later than 17500 BC already had the two features characteristic of today's hand sewing needle, including an eye at one end and a tapering point at the other end, and were made from materials such as bones and antlers. As people acquired metalworking skills, needles began to be made from metals such as copper during the Bronze Age in approximately 7000 BC and later from iron or bronze. Very few of these historical needles have survived due to corrosion; however, historical embroidery samples from this period indicate that the needles used to create this work were of a remarkably high standard. In 1755, a German named Weisenthal developed a two-point needle for use in early sewing machinery prototypes. Versions of this needle are used even today in modern industrial machines for sewing shank buttons or for imitating hand-made seams. The invention of the lockstitch sewing machine gave rise to the development of the sewing machine needle in the 1800s

DOI: 10.1201/9781003126454-2

where the design of the needle eye was moved close to the point. This eye-point needle paved the way for the mechanisation of sewing worldwide (Schmetz, 2005).

2.3 MORPHOLOGY OF THE INDUSTRIAL SEWING NEEDLE

The sewing machine needle is made from a variety of basic elements which can be arranged in different combinations to suit different end uses. These basic elements include the needle shank, the needle blade which can have one or two grooves, and the needle point with the eye and the scarf.

The *butt* of the needle is the flat end of the needle that facilitates insertion into the needle bar or clamp. The *needle shank* is the upper part of the needle which is normally wider than the rest of the needle, and it is this part of the needle that is attached to the machine needle-bar using either a screw or a clamp. The *blade* is the long section between the shoulder and the eye which has a *groove* on one side that runs down the length of the blade to provide a protective channel through which the needle thread is drawn in and out of the material during each stitch. The *scarf* is the indentation above the eye that allows the thread to be grabbed by the bobbin hook under the throat plate to create a stitch, and its size varies with different needle types. The *eye* located above the needle point is critical in reducing thread damage during penetration of the needle and in producing a good loop formation. The shape of the needle eye can also vary depending on end use, for example some have a larger cross section to reduce needle or fabric friction. The *tip* of the needle, in conjunction with the point of the needle, determines penetration performance and must therefore be

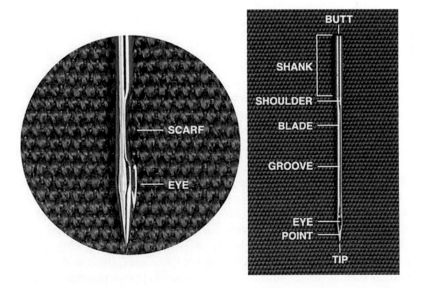

FIGURE 2.1   Anatomy of a sewing machine needle.

Source: Courtesy of www.Sailrite.com

chosen to suit the fabric being sewn. There are two basic classes of needle points, namely round point and cutting point.

2.3.1   ROUND POINT NEEDLES

The most common round point needles include Slim Set Point, Set Cloth Point, Heavy Set Point, Light Ball Point, Medium Ball Point, and Heavy Ball Point.

Set points are mainly used for sewing woven fabrics as they help to spread the yarn of the material apart without damage. Ball point needles are more commonly used for sewing knitted materials so that the rounded tip acts to deflect and enlarge the loops without bursting them.

2.3.2   CUTTING POINT NEEDLES

Cutting point needles have sharp tips but are available in a variety of cross-sectional shapes such as spear, wedge, triangular, or diamond and are used for sewing leather or similar dense non-fabric-based material. Cutting point needles pierce the material during the sewing action and therefore generate less needle heat. Sewability tests need to be undertaken when selecting a cutting point needle to ensure the desired strength and seam appearance.

2.4   MEASUREMENT OF SEWING NEEDLES: NUMBERING SYSTEMS

Sewing machine needles are measured in both the needle width and length. For the last 70 years, the numbering system for needle width has been globally standardised using a metric size categorisation known as Number Metric or Nm for short, which refers to the diameter of the needle blade, expressed as one-hundredth of a millimetre that is normally measured above the eye that goes through the fabric. The international standard ISO 8239 describes needle diameter as follows:

> blade diameter, **d**: Diameter at the cylindrical part of the needle blade above the short groove or scarf, but below any decrease or increase in cross-section of the blade. This diameter in millimetres multiplied by 100 corresponds to the metric size designation Nm = 90 designates a needle having a blade diameter of 0.9mm.

> (ISO 8239, 1987, p. 2)

The Nm number is normally stamped on the shank of the needle that is inserted into the sewing machine. Nm sizing generally ranges from 35 to 200, and the most widely used needles are Nm 60, 70, 80, 90, and 100 sizes. So, for example, an Nm 70 will normally be suitable for light sewing using 180- or 120-thread ticket size whilst an Nm 80 or 90 needle will be suitable for sewing heavy denim with a 30 or 25 thread ticket size. There are some exceptions to the latter where companies such as Singer have continued to use their own numbering system, but most of these companies also provide a conversion chart to assist accurate needle selection.

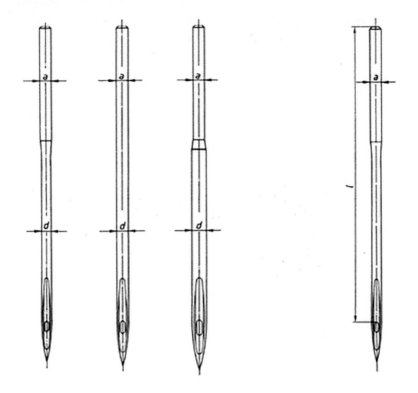

Figure 1 — Diameters of needles

Figure 2 — Length from butt
to eye of a needle

FIGURE 2.2 Measurement of sewing needles (ISO 8239, 1987, p. 2).

Source: Permission to reproduce extracts from British Standards is granted by BSI Standards Limited (BSI). No other use of this material is permitted. British Standards can be obtained from BSI Knowledge knowledge.bsigroup.com

The length of sewing machine needles is incredibly important because an incorrect needle length will usually result in seam or textile damage or even machine damage. According to ISO 8239 (1987), the length of a sewing machine needle is measured from the fixing end to the eye of the needle.

2.5 DESIGN OF THE NEEDLE FOR DIVERSE STITCHING TECHNOLOGIES

The sewability of a fabric is impacted by a myriad of factors such as complexity of design, fabric quality, fabric composition, and fabric finish such as wet processing and dyeing. Selecting the correct needle for each fabric or style can minimise production issues, such as skipped stitches, thread breakages, needle breakages, needle wear and tear, fabric and thermal damage, and so on, and help to minimise seam pucker and other quality issues such as pulled fabric threads. The following subsections will consider each of these sewing challenges in turn and demonstrate the vital role needle design selection and testing has for diverse stitching technologies.

2.5.1 SKIPPED STITCHES

Skipped stitches occur during stitch formation when the thread loop is not properly caught into the hook or looper, interrupting the interlocking or interlooping of the upper and lower threads. Skipped stitches impact the overall quality of a seam and therefore the integrity of the final product. In terms of sustainability, the impact of skipped stitches can mean that a whole garment can end up in landfill as seams unravel resulting in early disposal. The consequences of skipped stitches in other sectors such as the apparel performance sector can be serious especially where seam failure impacts performance and even health and well-being. Seam performance can therefore ultimately impact the reputation and overall success of a brand.

Most of the leading needle manufacturers in the world, such as Schmetz, Groz-Beckert, and Organ, have developed needles to minimise skipped stitches. This type of development involves the design of the scarf where in the case of the Schmetz SERV7, the needle is both reinforced by up to 20% and has a hump scarf which extends to the loop allowing the formation of a large loop of thread so that the hook or looper can catch it easily. The reinforcement of the blade makes the needle more stable when sewing thick fabrics or multiple layers and is less likely to bend which can also impact skipped stitching. This combination reduces or eliminates skipped stitches.

2.5.2 THERMAL MATERIAL DAMAGE

Thermal material damage can be evident when sewing materials with a high proportion of thermoplastic content and is caused when the sewing needle becomes overheated due to friction and results in the hot needle melting the fabric around the stitch hole. Melted residue can also adhere to the needle and can lead to fabric damage and thread breakages. Leading needle manufacturers such as Schmetz have developed needles with anti-adhesive coatings which eliminate the deposition of melted residue on the needle surface. The Schmetz NIT needle have a roughened phosphorated surface with a PTFE coating, meaning the needle stays cleaner for longer.

2.5.3 MECHANICAL FABRIC DAMAGE

Fabric damage occurs in fabrics that have a dense structure or those that have been treated with finishes such as starching or wrinkle resistance as the finishes often prevent the weave from being smoothly displaced when the needle penetrates the fabric and often results in the threads that make up the material that is being stitched becoming pierced by the needle rather than being displaced. The material threads can therefore become damaged or ruptured. Other causes of damage or rupture include the incorrect selection of sewing needle, particularly in relation to the thickness of the sewing needle and the design of the needle point. Ultimately, the thinnest sewing needle and thread should be selected where possible, and a rounded needle point can help to displace the weave thread during the stitching process.

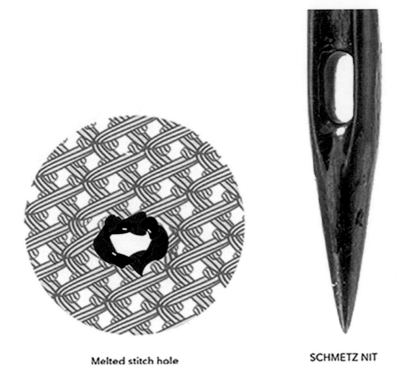

Melted stitch hole   SCHMETZ NIT

FIGURE 2.3   The Schmetz NIT Needle Ferd. SCHMETZ GmbH.

Source: Courtesy of www.schmetz.com

2.5.4   Knitted Fabric Damage

To understand how fabric damage occurs in knitted fabrics it is essential to explain the sewing of knitted fabrics. In simplest terms, when the needle enters the mesh structure it expands it to its fullest extent in the area of the eye. The mesh needs to draw on yarn from the surrounding material if it is to stretch to the required extent. Overstretching and tearing can occur if the needle is too large and the friction at the interloping parts of the meshes is too great.

2.5.5   Thread Breakage

There are a variety of reasons for thread breakages, such as slubs in the sewing thread, excessively high thread tension, mechanical damage, or synthetic thread breakages due to excessive needle temperatures. All these issues can be easily resolved by selecting better thread quality, adjusting machine thread tensions, or by applying synthetic thread lubrication devices.

However, other technical reasons for thread breakages include needle deflection away from the hook or looper which can also be corrected using highly stable sewing needles such as the Schmetz SERV7 range. According to Schmetz (2005), the size of the needle eye and the strength of the thread must match each other exactly so that the thread can pass through the eye with as little abrasion as possible.

2.5.6 SEAM PUCKERING

Seam pucker is visible as gathers along a seam and is an undesirable quality issue that can be prevented or minimised if factors such as stitch type, stitch density, stitch tension, feeding system, needle size, needle point shape, and sewing thread are carefully selected to complement each other. In terms of needle selection to minimise or eliminate seam pucker, according to Schmetz (2005), the needle and the thread should be as thin as possible to minimise displacement of the weave or knit by the sewing thread.

Another reason for seam pucker, particularly on synthetic fabrics, is due to the weave or knit structure being displaced and at the same time thermally set if the needle temperature becomes very hot. Once the displaced weave or knit is fixed in this position, even if the thread is removed from the seam the needle holes are unable to close, and the pucker remains as it is heat set in place. It is therefore important to carry out sewing tests involving material, needle, and thread ahead of manufacture.

2.5.7 PULLED FABRIC THREADS

Pulled fabric threads is mainly caused using needles that are too large in terms of both needle thickness and point size which can result in single threads of the material, that is being stitched, being drawn out of the weave structure by the tip of the needle. According to Schmetz (2005), in patterned materials, skipped stitches and interruption of the pattern repeat arise through pulled warp or weft threads or mesh threads.

2.6 KEY DRIVERS OF STITCHING NEEDLE ADVANCES

During the last 40 years there have been a variety of key drivers influencing sewing needle design and innovation. For example, during the 1990s at the height of globalisation involving the bulk production of apparel, the manufacturing sector relied heavily on technical solutions to minimise or eliminate any form of production downtime to achieve the greatest efficiencies possible. This may help to explain the high volume of new innovative needle designs that entered circulation from the 1990s onwards. These innovations included solutions for skipped stitches, needle breakage, thread breakage, and fabric damage. Even though all of these developments remain useful in today's apparel manufacturing industry, there are new challenges that are powering sewing needle design and innovation which include:

- The drive towards sustainable design practice
- Sustainable sewing needle supply chains
- Faster fashion
- Developments in new textiles, including technical textiles and trims
- Micromanufacturing for personalised and made-to-measure apparel
- Sewing automation

The drive towards sustainable design practice is particularly relevant to the garment sewing process at both the sampling and manufacturing stages. Innovative stitching needle design can be instrumental in reducing the volume of faulty apparel resulting from defective seaming and sewing damage, and this can reduce the volume of waste garments or damaged garment components generated during the manufacturing process.

Sustainable sewing needle supply chains are being driven by consumer demand for full product traceability as it has been recognised that there are aspects of industrial sewing needle design that can be instrumental in reducing environmental pollution. An example of this relates to the development of plating technologies that traditionally rely on chromium-based finishes, which can produce toxic effluents considered a high-risk environmental pollutant. Another major challenge for the needle industry relates to needle packaging. Traditional packaging solutions have long relied on single-use plastic because this has been proven to protect and preserve the quality of the needle from impact and humidity. Replacing single-use plastic with packaging solutions that are biocompatible, recyclable, and environmentally neutral has not been straightforward. There are other interesting developments in other sectors of needle design such as the knitting sector where award-winning needles such as the Groz-Beckert litespeed® needle have provided evidence that needle design can help in drastically reduce energy consumption during stitching and thereby reduce CO_2 emissions; and this provides further encouraging evidence that sewing needle design and innovation can be incredibly powerful and impactful where sustainability is concerned.

Whilst sustainability features high on most consumer's agenda, **faster fashion** involving the continual rapid production of smaller collections more frequently continues to flourish, and this necessitates the need to further improve needle design to combat traditional sewing manufacturing issues associated with production downtime, such as thread breakages, skipped stitches, fabric and seam damage, needle breakage, and so on.

Developments in new textiles including technical textiles have continued to accelerate in response to consumer demand for faster fashion and the growth in apparel markets such as shapewear, athleisure, workwear, and performance apparel. The diversity of the materials used across these markets can range from extremely fine microfibre cloths designed for intimate wear and sportswear applications to complex technical textiles used for the development of high-performance apparel with end uses such as Olympic performance, armed combat, and life-saving applications. Apparel manufacturers require sewing needles that can assure quality stitching regardless of base cloth construction, and this in turn drives the need for new innovations in sewing needle design.

Micromanufacturing, which enables the small-scale manufacture of luxury, or high-quality limited or bespoke apparel that is closer to market, is gaining momentum in response to consumer demand for personalised or built-to-last sustainable apparel. This further highlights the need for high-quality innovative sewing needles that can help to assure zero defects and stitching precision.

Sewing automation has existed in the apparel industry for over 70 years, but its relevance is gaining momentum particularly in relation to the machinery needed to support nearshoring, onshoring, and micromanufacturing but also in response to rising labour costs around the world. This driver is therefore influencing stitching needle innovations that can resolve automated sewing problems such as multi-directional sewing.

2.7 RECENT ADVANCES AND INNOVATIONS IN STITCHING NEEDLES

This section provides a summary of some of the key advances of industrial stitching needles used for stitching woven and knitted seams as well as the stitching of knitted structures.

2.7.1 THE SERV 7 NEEDLE

The SERV 7 needle has been developed by Schmetz to resolve issues associated with fabric damage and skip stitches which frequently arise when sewing elastic

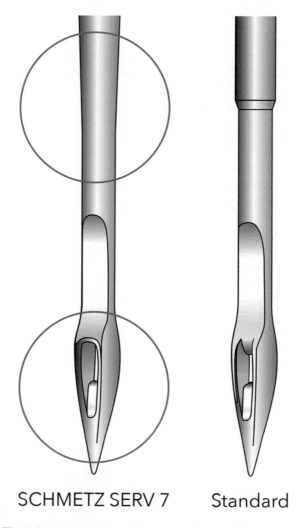

SCHMETZ SERV 7   Standard

FIGURE 2.4 The Schmetz SERV 7 needle in comparison to the standard needle.

Source: Courtesy of www.schmetz.com

material. The main distinguishing features of the SERV 7 design compared to a standard needle are the *hump scarf*, that makes a particularly large loop of thread which can be more easily caught by the point of the hook, and the *blade reinforcement*, which ensures the extraordinarily high stability of the needle. To produce the smallest possible stitch holes and optimise fabric displacement, the SERV 7 design often allows the use of a needle one size smaller without any reduction in the stability of the needle.

Figure 2.5 shows how the design of the SERV 7 forms a large loop of thread. The hump scarf extends the loop so that the hook or looper can catch it easily; this reduces skip stitches. In a standard needle, the loop is smaller and is therefore not caught by the point of the hook or looper and a skip stitch occurs.

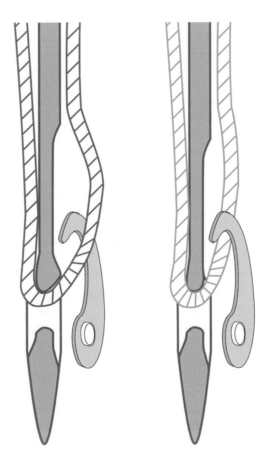

SCHMETZ SERV 7   Standard

FIGURE 2.5   The Schmetz SERV 7 needle supporting large loop formation in comparison to the size of the loop produced by a standard needle.

Source: Courtesy of www.schmetz.com

The SERV 7 needle also has a 15% reinforcement, making it highly stable and far less likely to bend, which can otherwise result in damage to the hook, looper, or other machine parts. The special stability of the SERV 7 needle is particularly valuable when sewing particularly thick materials involving multiple layers of fabric such as cross seams in jeans manufacture.

2.7.2   THE MR NEEDLE: FOR AUTOMATIC STITCHING APPLICATIONS

The volume of automatic sewing systems being utilised in the sewing industry is rapidly increasing in response to industry challenges that include rising labour costs, the growth of localised manufacturing, and the integration of robotics within the sewn product industries.

During the automatic stitching process, loop formation can often become unstable, especially when changing sewing direction. Standard needles frequently reach their limits in multi-directional applications, and this can result in sewing problems such as needle breakage, seam weakness due to skipped stitches or thread breakage, and damage to material. The MR needle (Groz-Beckert, 2022) has been developed to overcome these challenges. In terms of stability, the MR needle has an elevated level of deflection resistance due to its special blade and scarf geometry making it extremely resistant to bending. The MR needle has a deep and long-shaped scarf which facilitates tight adjustment of the looper to the needle. The deep thread groove which extends to the eye area promotes optimal protection of the thread, and the special asymmetrically shaped thread sliding area inside the eye of the MR needle guarantees stable loop formation even under unfavourable sewing conditions. This eliminates the possibility of unfavourable loop formation (*) and thread twist commonly associated with changing sewing direction.

2.7.3   THE GO NEEDLE: FOR THICKER DECORATIVE
STITCHING WITH SMALL NEED HOLES

This special needle has been designed by Schmetz for stitching situations where a thicker thread for decorative purposes is required but where a thicker needle is undesirable and unsuited to the base cloth that is to be stitched. This enables the creation of smaller stitch holes by using thinner needles without introducing the variety of thread problems normally associated with using heavy thread on fine needles.

The eye of GO needles is approximately 25% bigger, enlarged by two sizes in relation to the needle size, for example in a needle of size 70/10, the eye is of a needle size 90/14. Other key advantages include the use of thinner needles without changing the thread size, as well as the use of a thicker sewing thread without changing the needle size.

2.7.4   THE SF NEEDLE FOR EXTRA FINE KNITWEAR (40–42-GAUGE KNITWEAR)

The knitwear and intimate wear sectors have continued to expand in terms of market share but also in terms of the variety of products using extra fine knitted materials that can be incredibly difficult to sew without causing fabric damage or other stitching-related issues.

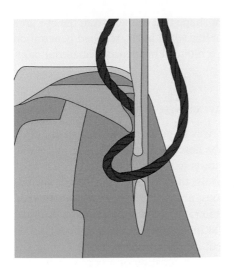
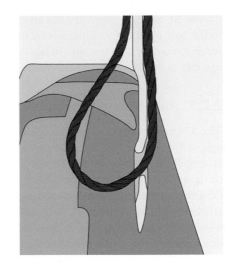

FIGURE 2.6   The special application MR needle.

Source: Courtesy of www.groz-beckert.com

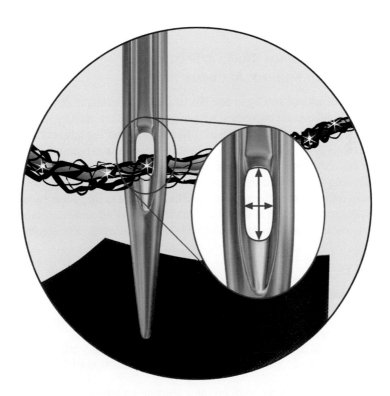

FIGURE 2.7   The Schmetz Go needle.

Source: Courtesy of www.schmetz.com

The SF needle's geometry has an extra slim shape that extends from the eye area to the ball point offering the advantage of smaller stitch holes, reduced penetration force in comparison to standard needles, and a reduced incidence of material damage due to the ultra-slim needle shape.

2.7.5   THE TN NEEDLE: SUSTAINABLE ANTI-WEAR NEEDLE

The TN needle, previously named SERV 100, was originally developed many years ago, but it is more relevant than ever and can be described as needle innovation that is supporting sustainable manufacturing due to its anti-wear properties. It has been designed for stitching complex abrasive materials, thick or coarse materials such as denim or leather, technical textiles, sports shoes, and synthetics; all of which when stitched with standard needles normally result in rapid wear and tear of the stitching needle that can lead to sewing defects such as skipped stitches, damaged panels, and downtime due to needle breakages and frequent needle replacement.

The TN needle has an ultra-hard titanium nitride coating on top of a protective hard chrome layer and a super hard tip, making it not only twice as hard as standard chrome-plated needles but also corrosion-resistant with additional anti-adhesive properties which is an ideal feature when stitching thick synthetic materials. These combined qualities enable an enhanced lifespan and therefore low needle consumption levels, high productivity, and overall reduced costs and waste.

2.7.6   THE NS NEEDLE FOR REDUCING SEAM PUCKER ON DELICATE SYNTHETIC MATERIALS

The NS needle developed by Organ needles in Japan has a unique design that helps to minimise or eliminate seam pucker on delicate synthetic materials (Organ, 2022). Many factors contribute to seam pucker such as differences in feeding layers of fabric which can slip and slide between themselves, tight thread tenson and humidity changes, and so on. The NS needle has a tapered needle point that is much slimmer than standard needles, allowing the needle to significantly reduce penetration resistance by up to 45%, which in turn results in the minimisation or elimination of seam pucker particularly on delicate synthetic fabrics such as those used in the design of athleisure or intimate wear. A similar needle has also been developed by Schmetz, called the KN needle.

REFERENCES

Groz-Beckert (2022) *MR Needle—Developed for Automated Sewing Processes with Multidirectional Feeding Systems* [online], https://portal.groz-beckert.com/germany/sewing/en/product-knowledge/special-application-needles/mr-needles,1142914 (Accessed 02/01/2023)

IOS (1987) *ISO 8239, Sewing Machine Needles: Fitting Dimensions Tolerances and Combinations* [online], www.iso.org/obp/ui/en/#iso:std:iso:8239:ed-1:v1:en (Accessed 02/01/2024)

Organ (2022) *NS Needle* [online], www.organneedle.com/img/pdf/trouble/Organ-Apparel.pdf (Accessed 02/01/2024)

Schmetz (2005) *The World of Sewing: Guide to Sewing Techniques*, Schmetz, Herzogenrath

3 Advances in Stitching Thread Technology

3.1 INTRODUCTION

This chapter will open with a brief overview of the variety of fibres used for manufacturing sewing threads as well as sewing thread types, thread construction, and thread size. The chapter will move on to consider special thread finishes such as lubricants, bonding finishes, anti-bacterial and anti-fungal finishes, water-repellent and anti-wicking finishes, as well as UV resistance. The chapter will also consider consumption calculations including digital advancements to support sewing thread consumption data, leading to a review of the latest advancements and innovations that are driving advancements in stitching threads, including the drive towards sustainable design practice, the growth in wearable technology and haptic apparel, developments in new textiles and technical textiles, sewing automation, on-demand manufacturing, and new models of sustainable manufacture to include micro-manufacturing. To provide the most current insight regarding recent and emerging advancements, key contributors include Coats Group Plc who are the world's largest manufacturer and distributor of sewing thread and A&E Gutermann Threads and Twine, a leader in on-demand thread dyeing sytems.

3.2 FIBRES USED FOR MANUFACTURING SEWING THREADS

Fibres used in the manufacture of sewing threads can be classified as either natural or synthetic. Threads made from natural fibres, which are affected by parameters such as moisture, rot, mildew, insects, and bacteria, are now outnumbered by synthetic fibres whose properties can include exceptionally high tenacity, high resistance to abrasion, and good resistance to fire and chemicals.

A summary of the main natural and manmade fibres used for the manufacture of industrial sewing threads used in sectors such as apparel, automotive, aerospace, and filtration are outlined as follows.

3.2.1 NATURAL FIBRES USED TO MANUFACTURE SEWING THREADS

Cotton is a natural fibre, and crops produce different grades. High-quality cotton sewing threads require cotton fibres that are strong and long but also fine with a typical diameter of 20 microns and a fibre length averaging approximately 38 mm. They can be treated to create either mercerised or glace cotton yarn with enhanced lustre and shine. Cotton is less common as an industrial sewing thread but is used due to its biodegradability, ease of dyeing, and excellent needle heat resistance.

DOI: 10.1201/9781003126454-3

Lyocell is made from regenerated wood pulp predominantly from sustainably sourced wood. It is an eco-friendly regenerated fibre which can be recycled and is fully biodegradable and compostable due to its cellulosic origin. In comparison to cotton, it does not require irrigation, pesticides, or labour-intensive processes during its cultivation. Coats threads have recently produced a Lyocell thread that is designed for apparel applications called EcoRegen, which is produced in a closed-loop manufacturing process that yields 99.7% solvent recovery.

Linen is one of the oldest textiles in the world, and the yarn is made from the stem of the flax plant. Flax is stronger, more absorbent but less extensible than cotton, which can be an advantage where 'no give' in the seam is required. Linen is easy to dye, has good fastness to UV light, and is considered an environment-friendly thread as it degrades naturally. Several of the large thread manufacturers produce linen thread for industrial application such as Coats linen thread for end uses that include automotive, bag manufacture, carpets, bedding, book binding, electrical cables, fishing lines, footwear, lace, medical supplies, sporting goods, and so on.

Fibreglass is considered a natural material as it is produced from sand. The latest development in fibreglass sewing thread includes a product called Glasmo. Coats Glasmo PTFE has been developed for the manufacture of demanding filtration applications. It is a temperature-resistant fibreglass sewing thread with exceptional thermal stability up to operating temperatures of 593°C. Fibreglass material does not burn, and E glass variety is used for spark plugs, thermo-couples, and ROHS-compliant colour-coded insulation wire assemblies. A variant of Glasmo QT is a Quartz sewing thread made from high-purity, continuous filament-fused silica with a maximum operating temperature of 2000°F/1094°C.

Natural **silk** thread is produced by silkworms when spinning their cocoons which are placed into boiling water to allow the cocoon to unravel so the silk filament can be extracted as silk threads. These fibres are used to create sewing thread, which is no longer used in the production of industrial sewing threads as it is considered to be unethical, unsustainable, and costly.

3.2.2   MAN-MADE FIBRES USED TO MANUFACTURE SEWING THREADS

Polyester thread is made from polyester raw material chips or granules (polyethylene terephthalate) which are melted and fed under pressure through a device known as spinneret, which is a metal disc with holes. The molten polymer is extruded through the spinneret to form continuous filaments. For the vast majority of sewing threads, the holes in the spinneret are circular, but for high-lustre filaments used to manufacture embroidery threads the spinneret profile is triangular, producing tri-lobal filaments. The diameter of the filaments is determined by the size of the holes in the spinneret and the pressure of the pump. These filaments are then cooled and collected together to form a continuous filament yarn. This is one of the main components for making polyester-based sewing threads. This process is known as melt spinning.

Polyamide (Nylon) 66 is produced in the same manner as polyester using adipic hexamethylenediamine, which produces the polymer hexamethylene diammonium adipate.

In addition to polyester and nylon, there are also a number of recent advancements in fibre technology that are used for the production of sewing threads. These are outlined as follows:

Polybutylene Terepthalate (PBT) is a type of polyester that is produced in the same manner as polyester. It is resistant to solvents, shrinks very little, is

FIGURE 3.1   Coats Lucense glow in the dark embroidery thread.

Source: Courtesy of Coats Group PLC

mechanically strong, and heat-resistant up to 150°C. Thread companies such as Coats have produced innovative threads from PBT yarns such as Coats Lucense, which is an embroidery thread with a special 'Glow in the dark' effect that can be 'charged' in sunlight or in artificial light and emits a luminous glow when viewed in the dark.

Homopolymeric Polyacrylonitrile is a polymer that is used to produce special-ised sewing threads that are acid- and alkali-resistant. Sewing threads made from this yarn such as Coats Dolanit are used for wet and dry filtration, that is for sewing filter bags and filter tubes which can be used for the separation of solid/gas mixtures and for the removal of dust from hot exhaust gases. Coats Dolanit is mainly used in cases where polyester is not suitable due to high moisture conditions.

PTFE (Polytetrafluoroethylene) is a synthetic fluoropolymer of tetrafluoroethyl-ene. It is hydrophobic, non-wetting, has high density, resistant to high temperatures and is well known for its non-stitch properties. Recent advancements include its use by Coats to create Helios-P, which is a highly specialised ultra-UV-resistant thread for outdoor use.

Aromatic Polyamide are more commonly known as aramid fibres which are a class of heat-resistant and strong synthetic fibres that are used in the manufacture of Kevlar and Nomex threads used for sewing Kevlar and Nomex fabrics with special fire-retardant qualities.

Liquid Crystal Polymer is used in highly demanding aerospace and military applications.

3.3   SEWING THREAD: TYPES, CONSTRUCTION, AND SIZE

3.3.1   THREAD TYPES

A wide range of sewing threads can be produced from the few fibres outlined earlier in the chapter and include:

- Core-spun thread
- Staple spun polyester thread
- Staple spun cotton thread
- Continuous filament threads

Core-spun Thread is the most common general-purpose sewing thread currently used in the apparel industry. It has two variants, namely Poly/Poly made from a polyester filament core with a polyester wrap and Poly/Cotton, which has a polyester filament core and a cotton wrap. The filament component is merged with the staple fibres during the yarn spinning operation. The filament takes up its position in the centre of the yarn with a protective sheath of staple fibres wrapped around it. These composite yarns are then twisted to form a plied thread (Coats, 2022c).

Staple Spun Polyester Threads (SSP) are manufactured from high-tenacity sta-ple fibres. A typical high-tenacity fibre used for sewing thread would be 1.2 denier, which is a measure of the linear density, and 38 mm long with a tenacity of at least 7.5 grams per decitex. Some thread manufacturers use a fibre length of 45 mm or even 55 mm depending on the machinery available. SSP threads are produced in a wide range of constructions and sizes, tex, and ticket numbers to accommodate most general sewing applications (Coats, 2022c).

Staple Spun Cotton Threads are made from high-grade long-staple fibres and include soft cotton, mercerised, and glace or polished. Soft threads receive no special treatment other than bleaching or dyeing and the application of a uniform, low-friction lubricant. Mercerised threads are treated under tension, in a solution of caustic soda which causes the fibres to swell and become rounder in the cross-section. This process enhances the lustre and increases the strength of the fibres. The dye affinity is also enhanced by this process. Glace cotton threads are produced from soft cotton threads by giving them a polishing treatment. This process involves the application of a coating of starch to the surface of the thread, brushing the fibre ends into the body of the thread, and drying them to form a smooth surface on the thread. The polishing process increases the strength of the thread by about 10%, but more importantly this process protects the thread from abrasion during heavy-duty sewing operations. Glace finishes are also applied to polyester and cotton core-spun threads (Coats, 2022c).

Continuous Filament Threads: Companies such as Coats produce a wide variety of continuous filament threads with different physical characteristics to satisfy particular sectors because they are significantly stronger than their equivalent size in core-spun, SSP (staple spun polyester), or cotton. Most are made from polyamide (nylon), polyester, and rayon, and the varieties include:

- Continuous filament thread
- Trilobal polyester thread
- Textured threads
- Locked filament polyester thread

Continuous Filament Threads: In the production of a continuous filament yarn, the filaments are gathered from the spinneret into a continuous strand, each strand comprising a specific number of filaments dependant on the desired characteristics. These strands are then combined and twisted conventionally into plied constructions similar to the post spinning processes used for spun threads. These threads are used in sewing applications where the seam strength is particularly important such as footwear and fine leather goods.

Some of these threads are given an additional treatment called bonding. In this case, the threads are coated with a soluble resin like nylon or polyurethane for continuous filament nylon and polyester or polyurethane for continuous filament polyester. After application, the resin is cured and dried, which has the effect of holding the plies together. The application of the bonding agents also reduces the abrasion on the thread during the sewing operation and is particularly useful for high-speed and automated sewing applications (Coats, 2022c).

Trilobal Polyester is a specific type of continuous filament polyester used for embroidery thread because it is modified to maximise the lustre of the thread created by the individual filaments having a triangular cross section (Coats, 2022c).

Textured Threads produced from continuous filament threads are designed to be bulkier and softer than the continuous filament twisted threads and can be produced in different ply constructions. The most common method of producing these threads is by false twisting. In this process, the filament yarn is subjected to heat, by contact or by convection, to soften the filaments. The yarn is then subjected to a rotational force which results in twist being inserted. As the thread exits the twisting zone of the machine, the twist is removed, but as the yarn is still in a thermoplastic condition the yarn's memory retains some of the distortion imposed in the twisting zone. This process results in the individual filaments adopting a crimped shape giving a soft bulky thread with high stretch characteristics. They are most suitable as under threads in lightweight chainstitching, overlocking, and cover seaming operations (Coats, 2022c).

Locked Filament Polyester sewing threads are produced by a technique involving the heating and stretching of the continuous filaments. The filaments are entangled and heat treated to produce a consolidated thread. Locked filament polyester threads have many of the characteristics and performance of threads produced by conventional spinning and twisting routes, but the thread does have a different handle or feel to it (Coats, 2022c).

3.3.2 SEWING THREAD CONSTRUCTION

All conventional sewing threads are constructed by spinning together relatively short fibres or twisting fine continuous filaments to produce basic yarns. Because of their fineness, these fibres and filaments have a large area of intimate contact with each other when held together with their axis parallel. This produces the coherence and strength combined with flexibility, which is essential in any good sewing thread, and it is the twist that is inserted in the basic yarns, usually in the 'S' direction, which produces the consolidating force. This is referred to as the *Singling Twist*.

The twist in the singling yarn is balanced by applying a twist in the opposite, usually 'Z', direction when typically, two, three, or four yarns are combined to form a sewing thread. This is referred to as the 'Finishing Twist'. Without the correct level of finishing twist, a conventional thread cannot be controlled during sewing. The individual plies can therefore separate during their repeated passages through the needle causing detriment to the sewing process.

Twist is therefore defined as the number of turns inserted per metre (or turns per inch) of the yarn or the thread produced. If the twist is too low, then the yarn may untwist, fray, and break; if it is too high, then the resulting liveliness in the thread may cause snarling, looping, knots, or spillage from the final package.

The term 'S' or 'Z' twist direction is derived from the diagonal of these letters following the direction of the twist. 'S' twist is sometimes referred to as right twist, and 'Z' twist is sometimes referred to as left twist.

The continuous filament yarns can also be twisted to make continuous filament sewing threads. Continuous filament yarns can also be assembled in two, three, or four plies to create continuous filament polyester or nylon sewing threads. These include textured threads.

3.3.3 SEWING THREAD SIZE: TICKET NUMBERING

Ticket numbering is a commercial numbering system used by thread manufacturers to reference the size of a given thread. The metric count, cotton count, and

Denier Systems use ticket numbering system to give an easy approximation of the specific size of the finished thread. It is notable that a ticket number in one type of thread will not be the same as in another, but in simple terms, the ticket number can denote:

- Higher the ticket number, finer the thread
- Lower the ticket number, thicker the thread

To assist with thread sizing and sewing needle selection, major thread manufacturers often produce apparel thread size conversion tables such as shown in Table 3.1.

3.4 SPECIAL THREAD FINISHES

Finishes are added to a thread to enhance sewability and function. The term sewability is used to describe a sewing thread's performance. According to COATS (2022b), a thread with good sewability is uniform in diameter with a good surface finish. Longitudinal uniformity of thread contributes to uniform strength and reduced friction as it passes through the stitch-forming mechanisms. It also minimises thread breakages and the associated costs incurred from rethreading machines, repairing stitches, and producing inferior quality products. All of these factors, and specifically thread strength, seam strength, and thread abrasion resistance, are directly impacted by the effectiveness of sewing thread lubrication.

TABLE 3.1
Apparel Thread Conversion Table

Tex Size	US Ticket	Metric Ticket	Cotton Count	Singer Needle*	Metric Needle*
18	120, 100/80	160	60/2	9–11	65–75
21	100	140	–	9–11	65–75
24	100, 100/60	120	–	10–11	70–75
27, 30	70, 70/40	100	60/3	12–16	80–100
35	70	80	–	12–16	80–100
40	50, 60/36	75	40/3	14–16	90–100
45, 50	40	60	–	14–18	90–110
60	30, T-60	50	20/2	18–21	110–130
80	20, T-80	36, 40	20/3	19–22	120–140
105	T-100	30	12/3	21–23	130–160
120	16	25	–	22–24	140–180
150	12	18	–	24–26	180–230

Note: *Needle size recommendations are nominal and ultimately depend on the sewing application.
Source: Coats (2022e)

3.4.1 LUBRICATION

Prior to the introduction of effective thread lubricants, compressed air blowers were often attached to sewing machines to cool the sewing needle. Developments in synthetic sewing threads and high-speed sewing from the 1970s onwards helped in highlighting the importance of thread lubrication research and development by major thread manufacturers which Tyler (2009) argues led to the advancements of cost-effective lubricants that did not stain or clog the needle eye whilst reducing friction and balancing slippage.

Thread lubrication is crucial for high-speed industrial sewing because the resistance and friction generated by the needle passing though fabric can result in the sewing thread being exposed to extremely elevated temperatures that can sometimes exceed 100 degrees above the melting point of polyester. This can result in thread abrasion, which can impact seam strength as well as causing thread and sewing needle breakages which can impact efficiencies as well as cause costly damage to machine parts. Thread lubrication can minimise these problems simply by cooling the needle. The most common thread lubricant is Silicone because it has excellent heat resistance and is usually applied post dyeing using the Lick Roller Method. Occasionally, additional lubrication may be needed, and this can be achieved by adding a thread lubrication device to the sewing machine. However, it is always recommended to test this process and to consult further guidance on suitability. Continuous innovations in thread lubrication technology have resolved most of the sewability issues caused by sewing friction and needle heat.

3.4.2 THREAD FINISHES

Thread finishes are also added to threads to enhance their performance and some of the key advancements in thread finishes include bonding, water-repellent anti-wicking, anti-fungal, fire retardancy, and UV resistance. Further details of these developments are elaborated as follows.

FIGURE 3.2 Sewing thread lubrication pot.

Source: Courtesy of www.sailrite.com

3.4.2.1   Bonding Finishes

Multi-directional and high-speed sewing creates thread rotation which causes traditional plied threads to untwist and fray. This can result in poor quality stitching appearance and compromised seam strength, a major risk factor when sewing high-performance apparel or safety equipment such as airbags for the automotive industry. Untwisting and fraying can also result in the loss of efficiencies due to thread breakages. This is particularly obvious with automated or semi-automated lockstitch sewing machine utilising stitching jigs. Several of the major thread manufacturers have responded to this issue by developing a variety of specialised bonded threads, which include, for example, A&E Gutterman ZWIBOND, a continuous filament thread for automatic sewing. AMANN has developed a bonded sewing thread called Strongbond for heavy-duty seams in the automotive and techtex sector. Strongbond is made from bonded Polyamide 6.6 continuous filament, which according to AMANN (2022) undergoes a bonding process known as gluing which provides a better cohesion of the thread's single yarn elements for multi-directional sewing applications because they are much harder to untwist.

3.4.2.2   Anti-bacterial and Anti-fungal Finishes

Bacteria and pathogens can often thrive around sew seams, and in response to this leading thread manufacturers have developed special sewing threads with anti-microbial and anti-fungal properties that prevent the growth of odour-and-stain-causing bacteria and pathogens around sewn seams. An example is Coats Epic Protect. This thread has been developed for use on a wide range of apparel, such as sports apparel, sports gloves, uniforms, including those developed by the military, workwear, as well as medical support and medical gowns. It has been stringently tested against the AATCC standards and passes a variety of fungi-resistance tests, which include no growth of mould culture *Aspergillus brasiliensis* (ATCC 9642), *Penicillium pinophilum* (ATCC 11797), *Chaetomium globosum* (ATCC 6205), *Gliocladium virens* (ATCC 9645), and *Aureobasidium pullulans* (ATCC 15233). It has also been confirmed that the finish is effective up to 100 laundry wash cycles (per ASTCC TM 150), meaning that the anti-microbial performance will endure for the lifetime of the product.

3.4.2.3   Fire-Retardant Finishes

Flame-retardant finishes are necessary for a wide range of apparel from industrial workwear for firefighting purposes to children's nightwear, which is governed by strict flammability legislation to include the Nightwear Safety Regulation (1985). This statutory regulation stipulates that sewing threads must meet the same flammability tests as the base fabric used to make the garment. Further details of the regulations are outlined by Tyler (2009), who states that 'synthetic nightwear materials should melt before they burn and extinguish themselves by the falling away of any burning, molten material' (Tyler, 2009, p. 119). Even non-flammable thread such as Nomex are not permitted for children's nightwear as Tyler (2009) notes, they can act as a wick and encourage burning of the garment material.

Demands for high-performance workwear have also assisted in driving new developments in threads that are manufactured from fire-retardant fibres such as Nomex as opposed to having specific thread finishes applied. Examples include Coats Firefly thread, made from 100% meta-aramid (Nomex), which is used for joining seams on firefighting apparel, industrial workwear, and military combat apparel. This thread provides protection from heat and flames up to 371°C. Other recent innovations include a wide range of ultra-high-performance threads from Gutterman such as Gutterman K-AR and Anesafe, used for joining protective workwear which can withstand temperatures up to 360°C. Guttermans's Aneguard (Kevlar) can withstand temperatures up to 490°C whilst their Acero thread is heat-resistant up to 720°C (under stress). All these Gutterman threads are flame-retardant, non-melting, and self-extinguishing.

3.4.2.4   Water-Repellent and Anti-wicking Finish

Water-repellent threads can be used for the design of showerproof apparel as opposed to waterproof apparel. It is notable that water-repellent finishes are not designed to make seams waterproof; this requires additional processes involving seam sealing and bonding or welding technology. Threads can be finished with a water-repellent finish to minimise moisture uptake when exposed to water as in rainwear seams on apparel or footwear. As well as reducing water ingress through such seams, water-repellent finishes can also reduce seam pucker, which can be caused by thread shrinkage when wet. Many waterproof sewing thread finishes in the past contained perfluorinated compounds (PFCs), which have been linked to climate change and are toxic. Recent development by leading thread manufacturers has created anti-wicking threads such as Coats Epic AWF that are PFC-free.

3.4.2.5   UV Resistance

UV degradation of sewing threads is concerned with the physical and chemical changes resulting from irradiation of the base polymers by UV or visible light. The energy absorbed leads to degradation of the polymeric chain and a loss of strength in the sewing thread. Moreover, it attacks the dyes concerned leading to fading or spotting of the thread colour. One of the most recent advancements in high-performance UV-resistant sewing threads is a product called Helios P, made from 100% polytetrafluoroethylene (PTFE), which is the strongest PTFE sewing thread made to date. Helios P is chemical-, heat- and UV-resistant, suitable for the production of outdoor goods such as awnings, boat covers, and boat sails that are exposed to harsh environments where UV rays, rain, and salt water can otherwise cause degradation. Other innovations in UV-resistant sewing threads include Guttermans WEATHERMAX®, Sunstop, Solbond, and Calora UV, all of which are designed to withstand extreme environmental influences of UV light.

3.5   SEWING THREAD CONSUMPTION

The calculation of sewing thread consumption was traditionally based on a simple calculation that involved measuring a specified distance on a seam and then unpicking and measuring it. Once this initial figure had been established it was possible to calculate the ratio of thread that would be used for all other seams on the product that used the same stitch and seam type. Due to variabilities in sewing methods and conditions such as thread breakages, reworks, differences in thread tensions, and so on, it was common practice to include a percentage of wastage to the initial calculation. This calculation would need to be repeated for every other seam/stitch combination on the garment, as different seams and stitch types use different proportions of thread. For example, a 301 lockstitch will use less thread than a class 500 overlock stitch, so it is important to select the correct ratio each time a sewing thread consumption is being calculated. Coats (2022d) provides a working example of this approach as follows:

Length of seam = 100 cm (1 m)
Stitch class 401 = 2-thread chainstitch
Length of seam for which thread is removed = 15 cm
Needle thread removed = 19.5 cm
Needle thread factor = 19.5/15 = 1.3
Looper thread removed = 62.0 cm
Looper thread factor = 62.0/15 = 4.1
Total needle thread = 100 cm × 1.3 = 130 cm
Total looper thread = 100 cm × 4.1 = 410 cm
Total thread consumed = 130 + 410 = 540 cm
Add 15% wastage = 540 cm × 1.15 = 621 cm

Many of the leading thread manufacturers such as Coats have developed thread consumption ratio charts such as the example presented in Table 3.2 that includes

TABLE 3.2
Thread Consumption Ratio Chart

Stitch Class	Description	Total Thread Usage (cm per cm of seam)/Thread Ratio	No. of Needles	Needle Thread %	Looper/Under (incl. Cover) Threads %
301	Lockstitch	2.5	1	50	50
101	Chainstitch	4.0	1	100	0
401	Two-Thread Chainstitch	5.5	1	25	75
304	Zigzag Lockstitch	7.0	1	50	50
503	Two-Thread Overedge Stitch	12.0	1	55	45
504	Three-Thread Overedge Stitch	14.0	1	20	80
512	Four-Thread Mock Safety Stitch	18.0	2	25	75
516	Five-Thread Safety Stitch	20.0	2	20	80
406	Three-Thread Covering Stitch	18.0	2	30	70
602	Four-Thread Covering Stitch	25.0	2	20	80
605	Five-Thread Covering Stitch	28.0	3	30	70

Source: Coats (2022d)

the different stitch types that are commonly used in the apparel industry. This makes the process of calculating thread consumption straightforward; a simple calculation can be undertaken that involves relating these predefined ratios to the length of seams and stitch type, plus a percentage for wastage which is normally around 15%.

More sophisticated methods of calculating thread consumption which support the fashion industry's ongoing digitalisation include the release of a new cloud-based thread consumption software developed by Coats Digital called SeamWorks 3. This software enables the accurate calculation of thread consumption and costing for a single sewn product or production run. SeamWorks 3 therefore supports the reduction of wastage as it is estimated that on average only 75% of all thread ordered ends up in the final sewn product, whereby the balance may be lost through operational wastage, through residual waste, or due to over ordering. This software therefore also supports cost-effective sewing thread procurement.

3.6   RECENT ADVANCES/INNOVATIONS IN SEWING THREAD TECHNOLOGIES

The sewing thread industry is well known for spawning key innovations that can enable remarkable products to be engineered and remarkable feats to take place. A key example is the development of Coats Atlantis thread, made from braided liquid crystal polymer (BRA), which was used to sew the airbag assembly that cushioned the landing of the Mars Pathfinder on the surface of the red planet.

FIGURE 3.3   SeamWorks 3 thread consumption calculator.

Source: Courtesy of Coats Group PLC

FIGURE 3.4 Airbag assembly that cushioned the landing of the Mars Pathfinder.

Source: Courtesy of Coats Group PLC

Recent advances and innovations in sewing thread technologies are currently being driven by key industry challenges and trends that include:

- The drive towards sustainable design practice
- Growth in wearable technology and haptic apparel
- Developments in new textiles and technical textiles
- Sewing automation
- On-demand, made-to-measure, and micromanufacturing

3.6.1 Sustainable Design

Sustainable design is being propelled by consumer demand for full product traceability and proven sustainable design practice. Thread companies have been at the forefront of sustainable innovations that enable the design and development of apparel that has a transparent end-of-life strategies designed into them.

Companies such as Coats, the world's largest manufacturer and distributor of threads, have recently made a public pledge that all their premium polyester threads will be made from 100% recycled materials by 2024 (Coats, 2022a). Also, they have recently launched a range of innovative sustainable threads that include EcoVerde, regarded as the first globally available 100% recycled range of premium core-spun and textured sewing threads that deliver the same proven level of performance as the industry's leading non-recycled threads. This thread range is considered innovative in other ways too, enabling apparel brands to design fully sustainable garments that are cut from and joined using recycled and sustainable materials *and* threads. Another advantage of this range of threads is that it has a significantly lower carbon footprint than virgin fibres. The EcoVerde range of threads are made from waste plastic that is collected from various industrial and post-consumer sources such as used plastic (PET) bottles which are sorted and cleaned and then ground to flakes. The flakes are melted down and extruded into fibre and filaments from which EcoVerde sewing threads are made. The range includes general sewing threads for lockstitch and overlockers, embroidery threads, and heavy-duty threads used for denim products.

To meet sustainability goals in combatting pollutants generated by crop farming and dyeing processes, Coats has launched two new threads that are 100% cotton called Tre Cheri Vero and Tre Cheri Vero Plus. Both have been certified by Better Cotton Initiative (BCI) as being made from sustainable raw materials and are completely free of harmful substances, including pesticides.

One of the most difficult issues that the clothing industry faces is finding solutions to enable efficiencies in supporting end-of-life cycle apparel recycling and the circular economy. Unpicking seams is a lengthy and labour-intensive process that is not cost-effective. Other approaches have included cutting the seams open, but this results in wastage and is still a labour-intensive process. Coats has recently launched a highly innovative thread called Ecocycle, the first water-dissolvable thread of its kind, that overcomes the challenges associated with garment recycling by dissolving at 95°C on a 30-minute wash cycle. After a seam dissolves, garments can be pulled apart into pieces with the easy removal of zips, buttons, and labels. Stringent tests have shown that this new innovation meets the same quality standards as virgin threads; it is conducive to ironing; and there is no effect on garment quality or integrity when washing at temperatures of up to 40°C.

Another interesting thread development from Coats is ECO B. This groundbreaking recycled polyester thread contains an innovative additive that reduces synthetic fibre accumulation in landfill and microfibre pollution in the oceans. This thread product marks a major step forward in sustainable design practice.

3.6.2 Growth in Wearable Technology Including Haptic Apparel

Key advancements in industrial sewing threads are supporting new developments in wearable technologies. One of the growth areas for wearable technology, and specifically wearable apparel, is the performance sportswear and gaming industry which commands huge revenues from the sale of apparel with integrated technology including haptic technology. For example, the Tesla suit, launched in 2019, was a full-body haptic suit aimed at the gaming world. It is constructed with 68 haptic points, capable of simulating a range of physical sensations all over your body. A limited range of these suits was produced in 2022 and auctioned using cryptocurrency, selling for US$20,000 each. Haptic technology has also been proven to enhance learning during training in performance sports such as speed skating whereby sensors in a smart suit can send real-time data to the trainer who can also send targeted instructions back to the athletes to correct, for example, a particular movement or stance whilst skating.

A key advancement in enabling the advancement of wearable technology has been the development of smart threads that can conduct electricity and data. A leading thread company called the AMANN Group has developed conductive threads that can be embroidered to create ultra-lightweight touch switches, and tests have shown they can replace traditional bulky plastic switches and cables. Further developments have shown that these smart threads can also be integrated directly into the fabric to function like an RFID antenna, and therefore record data, and this data may be useful for sustainability purposes such as recording the number of washes an item of clothing has had. According to AMANN (2022), conductive threads can also be suitable as pressure sensors that are used in composite material recording stress data, and this provides an insight into further development for the wider medical industries.

More recently, the Coats Group Plc has developed the Magellan range of futuristic 'smart' threads that can be integrated into the design of wearable apparel with capabilities for electro-conductivity, resistive heating, RFID technology etc. A summary of this range of smart threads include:

- *Electro-conductive*—Can be used to wirelessly charge, transfer power, carry messages, and actuate systems with a highly conductive material.
- *Resistive heating*—Can be used to pass current through a yarn and conductive material to heat items such as clothing, car seats, or even mattresses.
- *RFID technology*—Can be used to integrate a tag with an object so it can be automatically identified or assist with data capture for end uses such as smart textiles, automotive, health and body sensors.
- *Anti-static*—Can reduce the static charge build-up in fabric products used in clean room, assembly lines, and personal protective wear.
- *Electromagnetic shielding*—Used in the development of a barrier to electromagnetic radiation which can come from mobile phones, microwaves, or WiFi.

3.6.3 DEVELOPMENTS IN NEW TEXTILES INCLUDING TECHNICAL TEXTILES

Developments in new textiles including technical textiles have continued to accelerate in response to consumer demand for faster fashion and the growth in apparel markets, such as shapewear, athleisure, workwear, and performance apparel. The diversity of the materials used across these markets can range from extremely fine microfibre cloths designed for next-to-skin intimate wear and sportswear applications to complex technical textiles used for the development of high-performance apparel with end uses such as Olympic performance, armed combat, and life-saving applications. Apparel manufacturers require sewing threads that can assure quality stitching and seams regardless of base cloth construction, and this in turn is driving the need for new innovations in sewing threads.

Innovations in this area include a Coats thread called Seamsoft, which is a 100% recycled unique micro-filament-textured polyester thread that has been designed specifically for next-to-skin designs such as intimate wear and high-performance sportswear. Coats Eloflex is an innovative performance stretch sewing thread for high extension, enabling optimal stretch in a sewn seam. Its controlled elongation minimises the risk of seam extension failure on stretch fabrics. It has been developed in a range of thicknesses used for lighter-wear garments, such as football or rugby shirts, inner shorts, leggings, or even to help with the neck stretch measurements on childrenswear. It has a high initial modulus which enables it to assure good loop formation properties critical for good machine sewing.

Items of apparel such as combat trousers that have been designed using the highest standards of technical textiles can easily fail if a fastening such as a button falls off. To combat this, Coats has developed a fusible core-spun thread called Secura, which is a fusible polyester/polyester core-spun, self-locking thread offering an enhanced button security in the garments. The thread is impregnated with a fusible compound, which creates a fusible fibre matrix upon thermal activation by steam or normal iron and offers increased cohesion and enhanced button security. It is also an excellent product for safeguarding against choking on childrenswear and is certified to STANDARD 100 by OEKO-TEXe, class I, the most stringent class covering textile items for babies and toddlers.

3.6.4 SEWING AUTOMATION

Sewing automation has existed in the apparel industry for over 70 years, but its relevance is gaining momentum particularly in relation to the machinery needed to support micromanufacturing but also in response to rising labour costs around the world. This driver is therefore influencing sewing thread innovations that can support automated sewing problems associated with high-speed or multi-directional sewing.

Multi-directional and high-speed sewing creates thread rotation which causes traditional plied threads to untwist and fray. This can result in poor quality stitching appearance and compromised seam strength, a major risk factor when sewing high-performance apparel or safety equipment such as airbags for the automotive industry. Untwisting and fraying can also result in the loss of efficiencies due to thread breakages. This is particularly obvious with automated or semi-automated lockstitch sewing machine utilising stitching jigs. Several of the major thread manufacturers have responded to this issue by developing a variety of special threads, which include A&E Gutterman ZWIBOND, a continuous filament thread for automatic sewing (Guttermann, 2022). AMANN has also developed a bonded sewing thread called Strongbond for heavy-duty seams in the automotive and techtex sector. Strongbond is made from bonded Polyamide 6.6 continuous filament which according to AMANN (2022) undergoes a bonding process known as gluing, which provides a better cohesion of the threads' single yarn elements for multi-directional sewing applications because they are much harder to untwist.

Similar to A&E Gutterman and AMANN, Coats has produced a thread called Monobond, which is a multifilament monocord nylon sewing thread for twin-needle and high-speed multi-directional sewing. Coats Monobond is unique as it has a single-ply construction which eliminates ply untwisting problems. A single ply of multifilament nylon is lightly bonded using innovative 'tower-bonding' which means, unlike most other bonded threads, Monobond remains flexible and un-stiffened.

All the latter examples are aimed at the automotive or shoe industries; however, Coats has developed two threads aimed at the apparel industry to support multi-directional and high-speed sewing. The first is a bonded nylon thread called Nylbond Denim, aimed at the denim sector which commonly utilises semi-automated or automated machinery for sewing applications such as pocket setting, hemming, and waist banding. The thread is made from pre-stabilised continuous filament nylon 6.6. Unlike most other bonded threads in the market, this bonding approach produces a tough uniform bond with unique soft and supple construction needed for the modern stretch denim industry. It is also highly durable and can withstand parameters such as high stretch seam requirements or denim-based wet processes such as heavy bleach, stone, scraping, and other washing processes.

The second development by Coats aimed at the denim sector is an innovative thread called Dual Duty, which is made from 100% recycled material. Its high-tenacity core-spun polyester with a long staple cotton wrap also protects the core from needle heat, which can cause thread breakages, skipped stitches, yarn damage, and seam pucker. The filament core allows the use of finer thread sizes improving seam appearance but at the same time has the required strength and durability for high-speed stitching.

3.6.5 ON-DEMAND, MADE TO MEASURE, AND MICROMANUFACTURING

Trends in on-demand and made-to-measure apparel have recently re-emerged in response to increased consumer demand for personalised apparel. This expanding market has been evidenced by both online sales of faster smaller collections but also by a variety of new made-to-measure bespoke projects by companies such Levi's and H&M. This in turn has led retailers and manufacturers to re-imagine manufacturing such as micromanufacturing that can accommodate the latter. A key restraining factor in the apparel quick response or bespoke sector is speed of access to the right colour and quantity of sewing thread. A very recent advancement has successfully resolved this issue through the introduction of the world's first eco-friendly digital and waterless on-demand thread and yarn dyeing system called the TS-1800. This unique system has been developed and manufactured by a successful Israeli start-up company called Twine. The system can dye any of the shelf plain white polyester thread using the company's specially developed inks. The process starts with selecting a colour within the dedicated software, and this leads to the process of accurate ink mixing. The thread then passes through a treatment chamber and drying unit before a lubricant is applied. The entire system operates on a highly sustainable closed loop system, resulting in a vastly reduced carbon footprint, zero waste, increased efficiencies, and on-demand delivery.

Another recent development, this time aimed exclusively at the machine embroidery sector, is Coloreel which is an on-demand thread dyeing technology. The design is based on ink cartridges which contain ink, washing fluid, and lubrication that are mounted above each embroidery head. Thread is fed into the top of the cartridge.

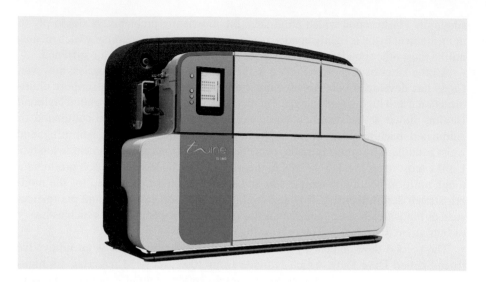

FIGURE 3.5 Twine's TS-1800 on-demand thread and yarn dyeing system.

Source: Courtesy of www.twine-s.com

FIGURE 3.6 Coloreel on-demand thread dyeing system.

Source: Courtesy of Coloreel.com

The Coloreel system dyes thread for each individual embroidery design. The benefits are similar to the TS-1800, except the system is sold with white thread exclusively produced and tested by Coloreel.

REFERENCES

AMANN (2022) *Strongbond Thread* [online], https://www.amann.com/products/product/strongbond/: (Accessed 20/02/2024)

Coats (2022a) *Sustainable Products* [online], www.coats.com/en-us/sustainability/sustainable-products (Accessed 02/01/2024)

Coats (2022b) *What Is Sewability* [online], www.coats.com/en/information-hub/selecting-your-sewing-threads#:~:text=%27Sewability%27%20of%20thread%20is%20a,through%20the%20stitch%20forming%20mechanisms (Accessed 02/01/2024)

Coats (2022c) *Thread Types* [online], https://www.coats.com/en/products: (Accessed 20/02/2024)

Coats (2022d) *Thread Consumption Guide* [online], https://www.coats.com/en/information-hub/thread-consumption-guide (Accessed 02/01/2024)

Coats (2022e) *Thread Conversion Chart* [online], www.coats.com/en/information-hub/thread-numbering#Count_Conversions (Accessed 29/12/2023)

Guttermann (2022) *Swibond Thread* [online], Sewing Thread for Airbag and Shoes and Leather from A&E Gütermann, Zwibond (https://industry.guetermann.com/en/products/zwibond/ (Accessed 20/02/2024)

Tyler, D (2009) *Carr and Latham's Technology of Clothing Manufacture*, 4th ed., Wiley-Blackwell, Oxford

4 Advances in No-Sew Seam Bonding and Ultrasonic Welding Technologies

4.1 INTRODUCTION

This chapter provides an update on a longitudinal study conducted by Mitchell and Hayes (2018) and Tyler et al. (2012) along with a comprehensive account of key advances and key drivers in no-sew joining technologies over the last five years. The original longitudinal study identified nine key drivers of this technology, which has since increased to ten, the most recent being sustainability. The other drivers of no-seam bonding and welding that are covered in this chapter include know-how/diffusion of knowledge, development of machinery and adhesives, cost and exclusivity, solution for defects when joining delicate fabrics, reduction in seam bulk/weight to enhance performance, waterproof performance, key sporting events, finishing seamless knitted apparel, and eliminate visible seam lines on intimate apparel. Consideration is given to how these advancements and drivers will inform future applications and markets, including new developments in associated machinery, equipment, and material. This qualitative research has engaged with several key manufacturers of no-sew machinery, equipment, trimmings, and adhesives; and key contributors include Tony Turner and Nick Turner (Sew Systems); Dave O'Leary, Wullie Daly (Ardmel Automation); Bill Reece (Macpi), and Dr Steven G. Hayes (University of Manchester).

4.2 ADVANCED SEAMING TECHNOLOGIES

Advanced seaming technologies, which utilise joining methods other than sewing with needle and thread, have enabled apparel designers to develop highly functional and innovative products. No-sew seam joining methods which include thermoplastic bonding, ultrasonic welding, and radio frequency welding have not only aided the creation of exclusive high fashion designer wear, but they have also supported and enhanced the comfort and performance of wearers engaged in professional competitive sports, as well as protecting a variety of end users from hazardous or life-threatening environmental conditions (Hayes and McLoughlin, 2008; Tyler et al., 2012).

DOI: 10.1201/9781003126454-4

4.2.1   No-Sew Technologies

Methods of joining textile materials to form garments other than sewing with needle and thread have existed for several decades. Commercial options essentially fall in to two categories, namely welding and bonding.

Growth in the thread-free seaming sector has been sporadic yet progressive and although previously restricted to specialised high-value niche sectors of the apparel industry, including intimate wear, close fitting active wear such as running or cycling, outdoor performance apparel, and workwear, has now diffused to all apparel sectors where its application remains positive due to the benefits afforded by improved comfort, aesthetic, and performance-enhancing streamlined fit, reduced bulk and weight, and enhanced waterproof performance.

Most garments that have been designed using no-sew technology are commonly produced using a combination of traditional sewing and bonding simply because many products or specific parts of products are either not required or designed to be joined using bonding or welding. Those items which utilise a high proportion of bonded or welded seams tend to be highly specialised items where elevated levels of comfort, performance, or protection are required or where the aesthetic seamless appearance is an essential feature of the design.

4.2.2   Welding

Initially, welding was applied to garment manufacturing most effectively in the production of short seams or stampings (such as vents, buttonholes, and eyelets). However, continuous welding can produce long seams and has seen recent acceptance as a method of joining garments. Several methods of welding exist; and in all methods the thermoplastic material is joined by raising the temperature of the substrate to such a level that melts the fabric at the interface of the two plies. Pressure applied in a controlled manner promotes the flow of one material into the other, and upon cooling a weld bead is formed. Welding techniques include hot air, hot wedge (thermal [resistance] sealers), dielectric welding (radio frequency [RF]), and ultrasonic welding.

Hot Air: A focused stream of heated compressed air is directed towards the join line of the interface between the fabric plies. A roller is placed directly behind this area to apply pressure once the material begins to flow and before or as it cools. The profile of this roller can define the weld impression.

Hot Wedge (Thermal Sealers): Electrically heated jaws are pressed into contact with the thermoplastic material for the duration required to transfer heat to the interface. The profile of the jaws dictates the weld impression.

Dielectric Welding (RF): Alternating waves from an electromagnetic generator excite the bipolar molecules of a dielectric material to such an extent that the molecular vibration causes friction between the molecules, which results in the fabric interface melting. During seaming, the fabrics are placed between two electrodes—one live electrode and the other a ground electrode.

Ultrasonic Welding: The two basic techniques are plunge welding and continuous welding. In plunge welding, the parts are placed under a tool or horn; the horn descends to the part under moderate pressure, and the weld cycle is initiated. In the continuous welding process, the horn may 'scan' the part, or the material is passed over or under the horn on a continual basis (Hayes, 2017; Tyler et al., 2012).

4.2.3   BONDING

Bonding uses an adhesive film (thermoplastic film) for bonding two fabrics (substrates) together. In this technique, the adhesive film is slit into tapes and applied in strategic locations, or the adhesive film can be laminated to wide-width fabrics. During the process, the adhesive is tacked or laminated to one of the substrates, and the second substrate can be laid on top. Then, heat and pressure activate the adhesive, the film melts and penetrates the fabrics and creates a bond between the two. This sequence of construction calls for different machines at each stage, the first for applying the film using the seam bonding tape attach machine (Figure 4.1) and the second for adding the additional layer of material and heating the laminate using the seam bonding ply adhesive machine (Figure 4.2).

FIGURE 4.1   Seam bonding tape attaching machine.

Source: Courtesy of www.sewsystems.com

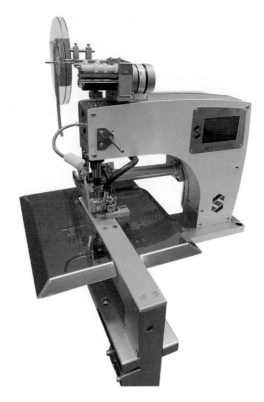

FIGURE 4.2   Seam bonding fabric ply adhesion.

Source: Courtesy of www.sewsystems.com

The nature of the bonding film, along with common variables of temperature, time, and pressure, dictates the performance of the join. Some films are more extensible than others, and a range of options are offered by suppliers such as Bemis, Ardmel, and Sew Systems. In fact, it is arguable that the acceptance of this technique for creating joins is linked very much to the recent developments in adhesive films (Hayes, 2017).

4.3   KEY DRIVERS AND ADVANCEMENTS OF NO-SEW INNOVATION

This section provides an update on the advancements and key drivers of no-sew joining technologies based on longitudinal study/research conducted by Mitchell and Hayes (2018) and Tyler et al. (2012). The original longitudinal study identified nine key drivers of this technology, which have since increased to ten, the most recent being sustainability. Recent advancements for each driver are outlined as follows.

4.3.1   DRIVER 1: KNOW-HOW/DIFFUSION OF KNOWLEDGE

Until recently, knowledge and awareness of no-sew joining technology methods remained poorly publicised and documented within the public domain. Existing information was dispersed and largely inaccessible within specialised textile technology textbooks, journals, and papers and was often in a technical language which was not easily understood by designers and therefore remained restricted and exclusive amongst those involved in the development of machines, adhesives, and methods. Manufacturers and suppliers were traditionally unwilling to give away valuable product development knowledge for fear of copying or competitive advantage, and therefore this knowledge often remained exclusive with large or wealthy manufacturers who were able to purchase knowledge, training, and support by investing in machinery and equipment. Smaller enterprises were often excluded due to economies of scale. More recently, the UK-based Sew Systems, who manufactures sew-free bonding and welding machines in the UK, has responded to this gap and have been running a training school to support new start-ups, helping them to develop products, expand, and become full adopters of this technology. Companies such as Ardmel, who specialises in the development of seam bonding and welding machinery and equipment and tapes have developed relationships with a variety of universities both in the UK and internationally, supporting students with research projects and dissertations that relate to advances in seam bonding and welding.

Diffusion of knowledge and know-how amongst young international designers and lifelong learners has also been supported by several universities such as the University of Manchester, De Montfort University, Albstadt-Sigmaringen University in Germany, and the Manchester Fashion Institute at Manchester Metropolitan University, who are one of the largest Fashion Schools in the UK whose heritage is steeped in fashion technology education. As such, no-sew technology is considered a standard item in their fashion curriculum, and their knowledge exchange department is also involved in supporting the fashion industry with research, education, and development in fields such as no-sew technology.

International brands such as Lululemon who specialise in no-sew have also recognised the benefit of educating the consumer by providing detailed online garment construction information helping them to make informed pre-purchase choices, specifically where performance or comfort is concerned.

The growth and development of this specialist sector have also been partly achieved because of highly established industry experts who possess a unique and expansive knowledge base needed to devise solutions for no-sew product development. This knowledge base involves a combination of no-sew technology complimented by an in-depth knowledge of traditional garment engineering and manufacturing technology. Many of these individuals have reached or are nearing retirement. It is therefore imperative that this experience is captured and passed on and that the fashion industry and higher education sectors collaborate to secure a concurrent stream of fashion graduates who can champion no-sew fashion design and technology for the future.

4.3.2   DRIVER 2: DEVELOPMENT OF MACHINERY AND ADHESIVES

Although aspects of no-sew technology have existed within other industry sectors since the 1940s (Herzer, 2005), uptake and growth within the apparel industry have been slow. This is partly explained by the difficulties associated with communicating the technology as outlined in driver 1 but is also due to the slow pace of the development of machinery, methods, and adhesives which has been largely demand based. For example, key brands such as Victoria Secrets and Marks and Spencer (M&S) inspired and accelerated demand at the beginning of the millennium, which in turn accelerated the development of machinery and adhesives.

More recently, brands involved in creating innovative no-sew apparel have worked in conjunction with machinery manufacturers to develop new and novel no-sew methods, and this in turn has accelerated the development of new machinery, equipment, methods, and adhesives. Many of these new methods are bespoke and 'built for purpose'. For example, a well-known active wear brand has recently developed a no-sew bra top where all seams, including highly shaped seams, are fully bonded. This necessitated the development of high-value bespoke machinery, with each machine costing around £60,000 (Figures 4.3 and 4.4). A spin-off benefit of the latter approach is the unlikely replication by competitors or copiers.

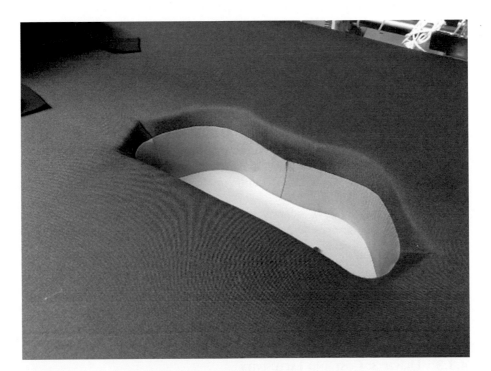

FIGURE 4.3   Bespoke MACPI pressing machine for bonding shaped components.

Source: Courtesy of Bill Reece

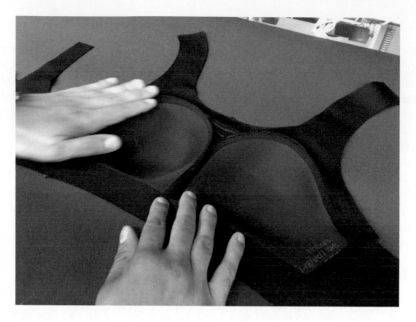

FIGURE 4.4   Shaped components inserted into pressing machine.

Source: Courtesy of Bill Reece

This *built-for-purpose* approach is also relevant to the development of adhesives, where for example highly specialised performance-enhancing overlay films used by brands such as Nike, Adidas, and Speedo are developed in collaboration with key adhesive manufacturers such as Bemis and Ardmel to meet the exacting requirements of a brand. In most cases, details of these bespoke *built-for-purpose* products remain highly confidential between the brand and the adhesive maker. This level of secrecy has meant that once a technology has been debuted, for example, at a key sporting event, it can be reused, or further developed for the future, or applied to mainstream apparel by the originating brand, thus creating income to offset the initial development costs.

Another example of how adhesives have accelerated the growth of no-sew apparel, specifically intimate wear and sportswear, relates to the development and improvement of high-modulus adhesives. Key advancements such as the introduction of Versafilm developed by Bemis in 2019 have enabled the development of 30% lighter and fully breathable waistbands due to the perforated structure of the adhesive material. Ardmel has also developed a specialised low-melt adhesive for stretch applications called T7950; it has 860% elongation and 98% recovery and provides excellent stretch and recovery for high-performance apparel and intimate wear.

4.3.3 Driver 3: Cost and Exclusivity

Regarding cost and exclusivity, it was originally thought that retail no-sew garments were approximately 10% more expensive than equivalents made using traditional sewing methods. However, this has proven to be a conservative estimate, and some manufacturers now agree that it can be up to 30% more expensive to manufacture quality no-sew apparel when compared to equivalents made using traditional sewing methods.

Reasons for the additional costs associated with developing and manufacturing no-sew apparel are outlined here, but ultimately the benefits enjoyed by the end user, which can include performance, comfort, or aesthetic appeal, appear to justify the higher cost in the eye of the consumer.

1. Investment in equipment and machinery is a costly risk which many companies wish to avoid, and therefore many products are sourced through specialised third-party manufacturers, who can charge a premium for product development and manufacture especially when operating within a monopolised environment.
2. Although trim costs are reduced due to the elimination of sewing thread, bonding adhesives and reinforcing tapes are still comparatively more expensive.
3. Manufacturing costs can also be increased because of joining defects. Unlike most stitched seams, if a bonded or welded seam is joined incorrectly, it is generally not possible to undo and re-weld or bond. Therefore, a whole garment may be wasted unlike most traditionally stitched items which can be reworked either by unpicking and re-stitching or where necessary, panel replacement.
4. Manufacture is slower in some cases as this is highly dependent on the equipment being used. Most rotary bonding and welding machines operate at much slower speeds than conventional sewing machines; therefore, throughput time is generally slower.
5. Most production orders tend to be smaller, therefore limitations on economies of scale usually lead to higher premiums. However, there are some situations where costs can be reduced, but this is dependent on product design. For example, it is possible to bond or weld several components together in one step, and many products can require fewer components. This is particularly relevant to the intimate apparel sector where, for example, an underwired bra which usually requires three stages of joining plus a variety of additional tapes and trims can be constructed in just one operation.

Exclusivity continues to drive this technology; however, a notable change regarding exclusivity that has taken place in the last ten years is that a much wider spectrum of apparel brands and sectors are now using this technology including high-volume value retailers such as Primark; mid-range fashion brands such as Zara; performance sportswear brands such as Nike, Adidas, Underarmour, and Canterbury; designer brands such as Tommy Hilfiger and Marithe Francois Girbaud; and the American brand Dyne who is well known for its innovative use of no-sew technology. Even the Italian Ministry of Defence has adopted this technology. However, the market is by no way saturated, and there are many new brands specialising in innovative wearable technology that are adopting this technology.

This widespread diffusion has been made possible, particularly at the budget end of the market, because the standard machinery, equipment, and adhesives, which were previously reserved for an exclusive customer base, have since been copied by vendors mainly located in the Far East, who themselves have ironically fallen victim to copying, for example, with machine engineers selling on information. There is little that can be done about this because it is complicated and expensive to pursue claims for copyright in offshore locations.

Therefore, at face value, in its basic form, no-sew is no longer an exclusive manufacturing technology. It's exclusivity and cost are determined by the way the technology is used to design and develop apparel which meets advanced requirements of the end user. These advanced requirements vary in degrees of intensity depending on the level of the market, which likewise dictates design and development costs. For example, Team GB's Lizzy Yarnold's skeleton suit which was constructed using no-sew technology cost £6.5 million to develop, test, and produce. The benefit here was it allegedly assisted this athlete to win gold at the PyeongChang 2018 Winter Olympics. At the other extreme, Primark, a European value retailer currently sells a variety of no-sew intimate wear, such as its three-pack 'invisible-hipsters'. These retail for as little as £10 but satisfy the needs of this type of consumers in terms of delivering at an affordable cost low profile seams, which are not visible when worn under outerwear apparel. It can therefore be argued that both products deliver exclusive benefits for the customer/end user.

4.3.4   Driver 4: Solution for Defects When Joining Delicate Fabrics

Developments in knitting and weaving technology have enabled the textile sector to produce extremely sheer and lightweight fabrics which are often high modulus, delicate, and prone to seam pucker and needle damage when joined using traditional methods using thread. Seam pucker can be caused by several variables, but the main cause of pucker when sewing fine fabrics is structural jamming. If the fabric has a dense construction, there may be insufficient space to accommodate a sewing thread without distorting the surrounding yarns. The common solution is to reduce needle size and thread size. However, this does not always solve the sewing problem, and attempts are usually made to press away pucker which often results in glazing, and the benefits in seam appearance are generally short-lived. Other common sewing problems include frequent needle breakages due to frictional heat stress and snagging. Therefore, thread-less bonding and welding offers a tangible alternative solution to the sewing of microfibre textiles, including those with high stretch and avoids any of the problems associated with needle and thread insertion. Likewise, most adhesives are chosen to have more stretch and recovery than the base fabric, and this eliminates other problems associated with seam failure due to seam cracking.

No-sew technology continues to offer a solution for manufacturers producing apparel using extremely sheer and delicate fabrics which are prone to pucker and needle damage. This is particularly relevant to the intimate wear, shapewear, and sportswear sectors where extra lightweight microfibre textiles are now commonplace. Orders involving highly delicate fabrics will often be subcontracted to third-party specialists who are highly experienced in this area, and most will also have additional experience using no-sew technology, which can be utilised when traditional joining methods with thread remain unsuccessful.

4.3.5   DRIVER 5: REDUCTION IN SEAM BULK/WEIGHT TO ENHANCE PERFORMANCE

When comparing seams joined with thread to those joined using no-sew technology, no-sew enables the construction of superior lighter weight and lower profile seams, which when worn next to the skin virtually eliminate the incidence of chaffing thus enhancing overall comfort as well as improved form-fitting. Both parameters are critical requirements when designing performance apparel for sports, such as golf, tennis, swimming, cycling, athletics, and so on. However, despite these advantages, including the fact that bonded and welded seams are as strong if not stronger than traditional sewn seams (Bemis, 2009), some prominent performance sportswear brands remain nervous about integrating fully bonded or welded seams into sports apparel where the end use involves contact such as shirt-grabbing. This is because there is a risk of the garment tearing or seams being pulled apart, which could result in disastrous press for both the branded supplier and the embarrassed wearer. This nervousness is partly based on historical events, for example in 2003, England's Rugby shirts were torn apart on the pitch after an episode of shirt-grabbing. It is notable that shirt tearing during rugby matches is commonplace; however, because these high-tech shirts had received a great deal of press coverage prior to the match, the media's reporting of the incident did little to clarify that it was not the technology that failed but the sheer brute strength involved in the contact shirt-grabbing during repetitive tackles. Even though this incident happened some time ago, brands producing football and rugby performance wear have continued to integrate no-sew into their products to achieve the low-profile seam, but most still tend to add a reinforcing row of threaded lockstitch as a security measure. More recently, companies such as Nike have surprisingly experimented with sewn seam technology for match-day football strips and have created less bulky seams using narrow bight overlocked seams with ultra-soft bulk threads that are then topstitched flat. This seaming method does produce a lower profile narrow non-bulky seam which although comfortable does not meet the same standards of form-fitting comfort as that afforded by no-sew seams.

4.3.6   DRIVER 6: WATERPROOF PERFORMANCE

High demand for waterproof apparel from a variety of sectors, which include Outdoor Pursuits and Performance Workwear, has assisted the development and application of bonding and welding technologies. For many years, apparel manufacturers of outdoor technical apparel have relied on traditional approaches involving a two-step process of stitching and hot-melt seam sealing using thermoplastic tapes to achieve waterproof seaming. These conventional methods were originally pioneered by Ardmel in the 1980s when it produced some of the first hot air seam-sealing machines. Other no-sew trail blazers include Framis in 1997 followed by Bemis and Arc'Teryx in 1998.

Since 2005, the evolution of more sophisticated ultrasonic welding machinery for joining apparel seams, such as the Ardmel Ultrasonic sewing machine, has offered an alternative solution for joining the wider range of thermoplastic materials which are now available. Likewise, advancements in thermoplastic adhesives have offered a solution for thread-less joining of non-synthetic fabrics which could not be welded together. Market demand for bonded and welded waterproof seam performance continues to be driven by sectors to include outdoor pursuits, outdoor sports, performance workwear, medical applications, and Ministry of Defence products. Throughout the recent global pandemic, the athleisure market propelled the use of bonding and welding technology to create fashionable and highly comfortable sportswear with functional features such as waterproof seams for everyday wear. According to Mintel (2020), unlike most other sectors of the fashion industry, the athleisure market remained buoyant through the global pandemic with over a third (32%) of people buying athleisure apparel to wear at home rather than for sports as demand for comfortable casual clothing increased as a consequence of COVID-19 lockdown measures.

During the global pandemic, the use of bonding and welding technology was unprecedented in the battle to create waterproof seams on medical-grade PPE apparel, particularly for the production of medical/isolation suits. Consequently, demand for seam sealing machines rocketed due to soaring orders across the world. The design of isolation suits is complicated, and each suit can take up to one hour to assemble. According to Dash (cited in Varshney, 2020), one of the biggest risks identified within this supply chain was the lack of understanding of the difference in seam sealing tapes and their use and application, which if selected or applied incorrectly, for example any wrinkles or creases can comprise seam integrity. This situation highlights the important role of higher education and those involved in the development of machinery and tapes to ensure the future generation of fashion graduates are fully conversant with the technical requirements of seam bonding and welding technology.

Leading adhesive developers and suppliers such as Framis, Bemis, and Ardmel have continued to develop and expand their range of innovative waterproofing and reinforcing tapes with features that include breathability, anti-cracking, abrasion resistance, and ultra-lightweight tapes that are virtually invisible and do not yellow. Tapes are also available with stretch properties that enable waterproof seam sealing or seam reinforcing around curves.

It will also be interesting to see how the glut of seam sealing machines that were built and deployed across the globe during this period will be used now that demand for medical PPE has subsided. Machine makers believe that the surplus machines might be used in two ways. First, towards the development of new advancements in waterproof products, and second, the industry is experiencing current shortages of electronic components, and many of the new machines will simply replace older technology where replacement parts are not available.

4.3.7   DRIVER 7: KEY SPORTING EVENTS

Key sporting events such as the Olympic Games, FIFA European and World Cup, the tennis Grand Slam tournaments, golf's Open Championships, and professional sailing races continue to drive the development of innovative apparel utilising no-sew technology. The performance-enhancing qualities afforded by this technology have been used to produce marginal gains in terms of speed and performance. However, split-second gains can make the difference in breaking a world record or winning an Olympic medal, and this has been the specific case for sports which make use of form-fitting athletic wear. Low-profile bonded seams are an integral part of skin-suit design. This is because bonded seams allow several different fabrics to be seamlessly zoned within a design by keeping the surface of skin suits as smooth as possible. This in turn helps to control airflow over different parts of the body, reducing friction, thus enhancing the aerodynamics of the design (Hart, J: cited in Reynolds, 2018).

Over the years, these advantages have sparked heated disputes particularly in the case of the skin suits worn by gold medal-winning Team GB's Cycling and Skeleton-race athletes during the 2016 Olympics and 2018 Winter Olympic Games where competitors claimed the suits provided an unfair advantage even though the International Bobsleigh and Skeleton Federation (IBSF) ultimately ruled they were legal.

Similar technology was also used in the suits worn by Team GB cyclists at the last three Olympics and took several years to develop. The technology, including the seam construction methods, remained highly guarded since first mooted in 2013, when the designs were filed as part of a published patent by UK Sport on behalf of engineers at the English Institute of Sport. Details of the patent show how ridges, involving 'kicks' shaped like chevrons, were bonded onto the back of the limbs to disrupt the flow of air around the arms and legs, thus reducing drag (Lewis, cited in Gray, 2013).

Developments for competitive sailing have included those produced by the brand Musto, an official clothing supplier for the well-known Volvo Ocean Race, which manufactured a fully bonded shirt for the 2014/2015 Volvo Ocean race (Figure 4.5).

Bonded seams were also a key feature of the 2016 Olympic and Paralympic swimming events, including the Adizero XV1, developed by Adidas, and worn by Paralympic champion Ellie Simmonds.

Many of these high-tech products have been redeveloped and produced in large volumes after key events, often marketed at high price points by specialist competitive sportswear retailers.

No-sew technology has continued to enable the development of highly technical skinsuits that have greatly improved athletes' performance at leading competitive events. Examples include Vorteq's Tokyo Edition skinsuit, which involved £400,000 worth of research and development over a three-year period and supported a variety of international athletes to win medals at the Tokyo 2021 Olympic games. Each suit was custom-fitted to the athlete using the latest 3D body-scanning technology along with 3D digital prototyping technology and wind tunnel verification to ensure the tension, fit, and no-sew seam placements were highly comfortable and fitted perfectly when seated in the riding position. Azizul Awang won gold and set a new national record at the New Zealand World Cup event.

The Huub Wattbike team also competed in the suit in the Brisbane World Cup event 2019, setting a new personal best time in their first competition with the Tokyo Edition Suit.

Growth in the use of wearable technologies for training purposes has further accelerated the use of no-sew technology which has enabled the seamless integration

FIGURE 4.5   Musto no-sew shirt.

Source: Author's own

of digital electronics into performance sportswear, thus eliminating bulky seams whilst providing a smooth comfortable insert. For example, during the last two Winter Olympics, many athletes trained with haptic suits that monitor body position during training, enabling trainers to recommend minor adjustments to their technique from a distance by vibrating different parts of the suit whilst in-wear.

There have been other notable developments relating to seam bonding and welding. For example, Bemis, a leader in bonding technology and Noble Biomaterials have developed a 'kit' for the creation of garments which are seamless and conductive. The kit includes a fabric which incorporates Noble Biomaterials' Circuitex technology. Fabrics incorporating Circuitex can be used to transmit a variety of data and conduct electricity. The kit also includes Bemis's High-Recovery no-sew adhesive and Bemis's Exoflex film which facilitate bonding during the production of seamless garments.

4.3.8   DRIVER 8: FINISHING SEAMLESS KNITTED APPAREL

In the past, garments created using circular knitting machines required a variety of edge-neatening finishes to complement their seam-free designs including straps on camisoles, waistbands on boxer shorts, edge bands on secret support shells, and hem turnings. Often, these finishes were achieved using seam bonding with thermoplastic adhesives, and this demand supported the development and use of innovative trims such as the 'Y' or 'V' elastic which are applied using bonding with thermoplastic adhesives. In the last ten years, concurrent developments in whole garment knitting mean that this fusion of technology is now rarely needed as knitting machines such as those produced by Santoni can make fully finished knitted seam-free apparel; whilst in the footwear industry high-tech knitting machines such as Shima Seiki enable the production of innovative 2D knitted uppers, originally pioneered by Nike in 2012 with the Nike Fly Knit Upper. These are knitted in 2D and then manipulated into a 3D shape and bonded onto a moulded sole.

FIGURE 4.6   Flat 2D Nike fly knit upper.

Source: Author's own

4.3.9   Driver 9: Eliminate Visible Seam Lines on Intimate Apparel

The development and use of no-sew technologies have continued to be driven by the multibillion-dollar global intimate and shapewear markets. According to Mintel (2022), the UK underwear, nightwear, and loungewear sector is expected to grow by 13% in the next five years to £5.5 billion by 2026.

Demand for no-sew underwear has remained buoyant since the introduction of the first sew-free bra, which was launched by Victoria Secrets in 2002 because this technology can produce non-bulky low-profile seams which facilitate a greater degree of comfort when worn next to the skin unlike traditionally stitched seams which can cause chaffing, discomfort, and visible lines on outerwear. Consumer demand for more comfortable no-sew underwear and shapewear has accelerated in line with the consumers' expanding waistline. According to the Health Survey for England, 64% of all adults were classified as overweight and obese in 2019, and according to Mintel (2022), this trend has led to increased demand for comfortable no-sew plus-size products such as minimiser bras and shapewear across all age groups Technavio (2018).

According to Mintel (2022), there has been a rising popularity in sports bras and non-wired crop top bras amongst Gen Z and younger Millennial females. This further highlights the shift towards more comfortable styles as people spend more time at home or undertaking a fitness-related activity. This accelerated growth is also creating competition between brands which according to Technavio (2018) is encouraging many of them to enhance their intimate wear product portfolios by introducing innovative and advanced no-sew products. This is particularly the case amongst mid-market clothing retailers such as H&M, the supermarkets, and value retailers such as Primark, all of whom have established no-sew products within their lingerie portfolios.

FIGURE 4.7   H&M bonded bra top.

Source: Courtesy of www.hm.com

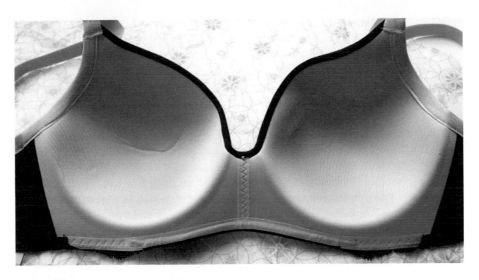

FIGURE 4.8   Triumph Bra with integrated 'Secret Wire'.

Source: Author's own

According to Technavio (2018), growth in the use of sew-free technology within the lingerie and shapewear sector has also been driven by the trend of consumers wanting to purchase shapewear that matches the colours and patterns of their daily clothes. This trend has been facilitated by the development of coloured adhesive films and tapes which are needed when bonding different fabric colours.

Likewise, improvements in laminating technology which enable fabric plies to be seamlessly bonded together have enabled a variety of shapewear products to be developed, such as uplifting butt-enhancing pants, which involves bonding a cushioning mid layer between top and under fabric plies.

Other innovations in this area include the innovative Magic Wire© originally developed by Triumph in 2014, which enabled flexible silicone wires to be laminated/bonded into a bra (Figure 4.8) thus replacing uncomfortable metal wires. This development has become standard in many moulded bras.

A long-standing criticism of no-sew underwear is that it has tended to look very plain, specifically when compared to traditional lingerie products. Developments such as flocked and textured overlay films, along with ultrasonic stitching wheels which create lace-look patterns and scalloped edges have assisted designers to create no-sew intimate wear products with an improved feminine aesthetic whilst maintaining an anti-visible seam line appearance.

4.3.10   DRIVER 10: SUSTAINABLE NO-SEW TECHNOLOGY

The most recent driver of no-sew bonding technology relates to the sustainability of materials used in the construction of bonded or welded seams. Two of the leading thermo-adhesive manufacturers Bemis and Framis have recently developed sustainable adhesives.

Bemis Associates Inc. has partnered with a company called Novoloop in 2021 to pioneer the first TPU made from post-consumer polyethylene waste that matches the performance characteristics of virgin TPUs made from petrochemicals. The new product, called Oistre™ (OYST-rah), is a thermoplastic polyurethane (TPU) for use in high-performance applications such as footwear, apparel, and sporting goods. According to Husband (2022), Oistre's carbon footprint is at least 46% smaller than conventional TPUs, and this development marks a major shift away from the use of virgin petroleum for TPU production.

Bemis has also partnered with DuPont Tate & Lyle Bio Products to develop its first Sewfree® product made using renewable material. According to DuPont (2021), bio-based Sewfree® 3700 film has been developed utilising 25% bio-based Susterra® propanediol from DuPont Tate & Lyle Bio Products, which is renewably sourced, plant-based, and has been engineered to replace elastics and bulky sewn seams in garments. The new product also meets the Apparel and Footwear International RSL Management (AFIRM) group Restricted Substance List.

Framis, a leading manufacturer of thermoplastic adhesives and thermo-adhesive bonding machines, has launched a water-based and solvent-free tape and film, called Rio, designed to reduce the environmental impact of both product and process. According to Framis (2022), to make RIO, the polyurethane resin in DMF (dimethylformamide) is replaced by dispersion of polyurethane in water, meaning the traditional solvent abatement process that was previously used can be eliminated, meaning they can now create a solvent-free product.

Other companies who continually strive to meet the highest standards of sustainable practice include Ardmel, who has introduced recycled release paper into their production process when manufacturing adhesives. The paper is used as a backing material during extrusion.

Different solutions are continually being sought, shared, and used to support sustainable practice within the seam bonding and welding sector.

4.4 CONCLUSIONS

To conclude, there have been many positive changes and developments regarding the no-sew technology sector. Its growth has been accelerated by several factors including an array of educational initiatives supported by key machine and adhesive makers, universities, and a number of apparel brands that have realised the financial advantages of educating their customers about the technology behind their clothing products.

Cost and exclusivity are still driving this technology, but these parameters are also influenced by other factors to include market diversification and the replication (copying) of machinery and adhesives. At face value, in its basic form, no-sew is no longer an exclusive manufacturing technology. Its exclusivity and cost are determined by the way the technology is utilised to design and develop apparel which meets advanced requirements for the end user, who is now as diverse as an Olympic athlete or a value-driven customer seeking affordable intimate wear at retailers such as H&M or Primark. The latter products will often utilise standard no-sew machinery and adhesives, the former will often engage in a 'built for purpose' approach, whereby new machinery and adhesives are developed to the bespoke requirements of a brand, usually operating at the top end of the designer of performance sportswear markets. This approach enables brands of ultra-high performance to develop distinctive innovative products. Such developments are costly, and brands have further realised the need to keep their innovations confidential, giving them the opportunity to reuse or further develop for the future. For many, particularly in the sportswear sector, if the technology can be kept secret for long enough, a way of recouping 'built for purpose' costs is to eventually integrate it into a mainstream mass-produced product. Such approaches have therefore facilitated the development of new no-sew machines, adhesives, and methods, which provide innovative solutions to achieving qualities such as breathability, insulation, compression, anti-muscle vibration, waterproofness, sustainability, or those that enable construction methods to be simplified. And this has made the technology much more attractive to a wider audience of designers than ever before.

No-sew technology continues to offer a solution for joining extremely sheer and delicate fabrics which are prone to pucker and needle damage. However, such orders will often be subcontracted to third-party manufacturers who specialise in no-sew manufacturing.

Market demand for bonded or welded waterproof seam performance continues to be driven by sectors including outdoor pursuits, performance workwear, Ministry of Defence products, and more recently the COVID-19 global pandemic, which necessitated the unprecedented mass manufacture of PPE apparel. The pandemic has also further elevated the growth of athleisure apparel and in particular athleisure with waterproof seams as many consumers have continued with the outdoor exercise prescribed by the government during the pandemic.

Growth in the use of wearable technologies has further accelerated the use of no-sew technology by enabling the seamless integration of digital electronics into a wide variety of apparel.

Both lower profile seams and the integration of wearable technologies have assisted athletes to achieve marginal gains in terms of speed and performance during official events and races. Such advantages have continued to spark heated disputes and claims of unfair advantage, which are commonly reported by the global media, and ironically, this has helped to intensify interest and uptake of the technology. However, media reporting has also negatively affected how the technology is used in sports which involve contact or shirt-grabbing, and despite ongoing developments in no-sew technology, some leading brands still reserve no-sew technology for non-contact performance sports.

Regarding the impact of no-sew technology within the seamless knitting sector, in the last ten years the focus has shifted away from apparel to footwear which has focused on innovations associated with knitted uppers bonded to highly technical moulded soles.

The multibillion-dollar global intimate and shapewear market also continues to drive no-sew technology. Consumer demand for more comfortable underwear and shapewear has accelerated in line with the increase in obesity. This accelerated growth in intimate wear is also creating competition particularly between value apparel retailers such as H&M and Primark to compete by introducing innovative and advanced products, involving seamless innovations. This driving force has provided the incentive for adhesive makers to develop high-modulus or coloured adhesive films needed when bonding different fabric colours. Likewise, improvements in laminating technology have enabled a plethora of highly successful no-sew shapewear products and innovations to repeatedly flood the market. The long-standing criticism of no-sew lingerie looking too plain and lacking a more feminine aesthetic has been conquered through the creation of faux lace and scalloped edges created using ultrasonic tools or the addition of bonded motifs and textured transfers.

The future of no-sew looks healthy. In the short term, the key drivers as discussed earlier look to continue to support growth and development of no-sew technology. For the medium term, several developing markets including Athleisure and the wider wearables market will potentially help the no-sew industry to expand and develop further.

Consumer pressure is forcing the apparel industry to change and adopt meaningful sustainable practice, and the so-free community is responding with innovative developments of sustainable thermo-adhesives by leaders such as Framis and Bemis. The sustainability element may therefore be seen as a new driver of this technology.

In the longer term, as no-sew technology matures, new markets such as the supermarket brands may employ this technology as a way of producing higher quality, distinctive value-added apparel which can compete against many of the high-street brands. It is unlikely no-sew will ever fully replace threaded joining techniques; however, this is a successful technology whose metamorphosis is far from complete.

REFERENCES

BEMIS (2009) *Is Sewing Really Out of Date?* [online], www.bemisworldwide.com/resources/ (Accessed 03/01/2024)

DuPont (2021) BEMIS Associates Partners with DuPont Tate & Lyle Bio Products to Launch First Sewfree® Product Made from Renewable Material, *BEMIS* [online], www.bemisworldwide.com/bemis-dupont-tate-lyle-bio-products-launch-first-sewfree-renewable-material/ (Accessed 29/12/2023)

Framis (2022) *Rio Is Born* [online], https://framis.it/en/rio-the-first-water-based-and-solvent-free-tape-and-film/ (Accessed 02/01/2024)

Gray, R (2013) British Cycling Team Develop New Drag Resistant Clothing and Helmet. *The Telegraph* [online], www.telegraph.co.uk/sport/othersports/cycling/10461757/British-cycling-team-develop-new-drag-resistant-clothing-and-helmet.html (Accessed 02/01/2024)

Hayes, SG (2017) *Joining Techniques for High-Performance Apparel*, Elsevier: The Textile Institute Book Series in Association with Woodhead Publishing, Cambridge

Hayes, SG and McLoughlin, J (2008) Chapter 10 – Technological Advances in Sewing Garments. In: Fairhurst, C (ed.) *Advances in Apparel Production*, Woodhead, Cambridge

Herzer, K (2005) Welding: A Tradition with a Future. *JSN International*: 13–16

Husband, L (2022) *Upcycling Start-Up Raises Funds to Convert Plastic Waste into Apparel* [online], www.just-style.com/news/upcycling-start-up-raises-11m-to-convert-plastic-waste-into-apparel-materials/?cf-view&cf-closed (Accessed 02/01/2024)

Mintel (2020) *Sports Fashion: Inc Impact of COVID-19—UK—December 2020* [online], https://reports-mintel (Accessed 03/01/2024)

Mintel (2022) *Underwear: MINTEL Market Research Reports* [online], https://reports-mintel-com.mmu.idm.oclc.org/display/1097111/?fromSearch=%3Ffreetext%3Dunderwear%2 5202022%26resultPosition%3D1 (Accessed 03/01/2024)

Mitchell, A and Hayes, SG (2018) *The No-Sew Revolution—The Next Chapter, The 91st Textile Institute World Conference, Leeds* [online], www.textileinstitute.org/wp-content/ uploads/2021/03/TIWC-2018-Programme-Final.pdf (Accessed 11/12/2023)

Reynolds, M (2018) *How Skeleton Skinsuits Gave Team GB the Edge at the Winter Olympics* [online], www.wired.co.uk/article/skeleton-suits-team-gb-winter-olympics-lizzy-yarnold-dom-parsons (Accessed 02/01/2024)

Technavio (2018) *Global Intimate Apparel Market 2018–2022* [online], www.businesswire. com/news/home/20180706005362/en/Global-Intimate-Apparel-Market-2018-2022-Key-Factors-Driving-Market-Growth-Technavio (Accessed 02/01/2024)

Tyler, D, et al. (2012) Recent Advances in Garment Manufacturing Technology: Joining Techniques, 3D Body Scanning and Garment Design. In: Shishoo, R (2012) (Ed) *The Global Textile and Clothing Industry*, Technological Advances and Future Challenges, Elsevier, Cambridge, pp. 131–170

Varshney, N (2020) *What Goes into Making a Body Coverall* [online], https://apparelre-sources.com/technology-news/manufacturing-tech/goes-making-body-coverall/ (Accessed 02/01/2024)

5 Automation and Robotics in the Apparel Manufacturing Industry

5.1 INTRODUCTION

This chapter will focus on automation and robotics in the garment manufacturing industry for cutting, spreading, and sewing. To contextualise the development of automation in the garment manufacturing industry, the chapter will open with a brief history of automation and industrial change from the 1950s and concurrent advancement in mechanised, semi-automated, and automated sewing machine systems and automated cutting and spreading technology. The chapter will consider the challenges of automating interchangeable designs and non-standard materials as well as the current drivers of automation in the apparel industry, including increasing labour costs, pressure from consumers to meet sustainable manufacture through localising manufacture, on-demand manufacture plus the positive developments in computer hardware, software, AI, and robots that have also acted as enablers of automation. The chapter will move on to provide a review of the latest advances in sewing, cutting and spreading automation and will finally consider the future of automation in the apparel manufacturing industry. To provide the most current insights regarding recent and emerging advancements, key contributors include representatives from PFAFF, JUKI, Durkoff Adler, Yamato, Mitsubishi, and Vi BE Mac, who were interviewed at the Texprocess industrial machinery exhibition in Frankfurt in June 2022, and also Jonas Hillenburg Vention.

5.2 WHAT IS AUTOMATION?

Some of the most successful and profitable footwear and apparel manufacturers in the world have a commonality which is based on their ongoing investments in automated technologies that have transformed the way they work. For companies such as Hugo Boss, Levi's, and Nike, this has dramatically further improved their success, their profit margin, their impact on sustainability, and their reputation (McKinsey, 2018b). For example, Hugo Boss has demonstrated how Industry 4.0 works in practice by transforming its largest manufacturing site in Izmir Turkey into a high-tech smart factory where employees work alongside highly automated machines and robots, which are collectively networked with the ability to analyse data and optimise workflows to achieve the highest efficiencies possible. Levi's has introduced an automated laser solution for finishing its jeans that shortens the process from 20 minutes to 90 seconds which means it can nearshore or onshore unfinished jeans, which can be quickly finished based on customer demand and rapidly delivered to retail shops within days

DOI: 10.1201/9781003126454-5

with no overstock. Nike has taken the bravest step of all and experimented with the onshoring of fully automated high-performance running shoe manufacture at two manufacturing sites in Germany and the United States called SPEEDFACTORY by substituting human labour with robotic manufacturing. The project which terminated in 2022 has provided essential learning for future advanced automation for footwear and clothing manufacture.

According to IBM (2022), automation is a term for technology applications where human input is minimised. A fully automated manufacturing supply chain is one where a product is manufactured from start to finish with no human intervention by employing automated machinery and robots working in tandem.

This amplified and necessary movement towards automation in manufacture can be conceptualised as the Fourth Industrial Revolution, also known as Industry 4.0. In its simplest form it can be described as the blurring of the physical and digital through the joining of technologies such as artificial intelligence (AI) and both mechanical and robotic automation. A vital aspect of this joining of technologies is that they can also be connected, controlled, and therefore driven without human intervention using smart technologies and AI-assisted software, which can monitor safety, fault detection, predictive maintenance, or the tracking of productivity (Ghobakhloo, 2020). Automation can also improve efficiencies in the areas of production planning and sequencing using AI-based software such as Robotics Process Automation (RPA), which is a special AI-based software that is designed to automate high-volume and repetitive rule-based tasks to enhance productivity and overall efficiencies. An example of its usefulness is in its use for planning and sequencing factory orders. Traditionally new orders are often received at the end of the day, meaning the planning and sequencing will often be undertaken by the factory manager or deputy the following morning when the working day resumes. However, with the RPA software, this can be done immediately once the order is received, without the need for human intervention, therefore enhancing planning and productivity (Varshney, 2021).

5.3 POTTED HISTORY OF AUTOMATION AND INDUSTRIAL CHANGE SINCE 1950

The journey of automation in apparel manufacturing dates back to the invention of the mechanical lockstitch sewing machine in the 1840s which enabled clothing to be constructed quickly and more efficiently than labour-intensive hand methods. The success of this initial invention stimulated the innovation of a plethora of other types of mechanical sewing machine innovations during the period 1850–1890. A notable development in this period was the four-motion-drop feed, which is a key feature on most modern sewing machines. Other key examples that successfully replaced labour-intensive hand sewing methods included the merrow overlock stitch in 1868, the zigzag stitch in 1873, chainstitch 1857, and the Reece buttonhole in 1880. Forty years later, the first electric sewing machines were introduced transforming civilian clothing and military uniform manufacture and supply during both world wars.

Manufacturing apparel in the Western world using local home-grown labour and mechanical methods remained cost-effective up until the 1950s, when rising labour costs and cheaper imports from the East began to progressively threaten the future of the apparel manufacturing industries in the West. Even at this early stage, key machine developers PFAFF recognised the importance of automating sewing production during the 1960s, resulting in the Transfer Street System, which enabled the production of a shirt front in sequence of manufacturing operations. Although it had many limitations and was at this time commercially unviable, many of its key features assisted in paving the way for the development of the wide range of semi-automated workstations that were developed from the 1980s onwards (Mcloughlin and Mitchell, 2013).

According to Jones (2006, p. 25) 'the period between 1978 and 1983 marked a watershed followed by a steady decline of employment in the UK apparel industry'. The crisis in the UK apparel industry at that time was directly attributed to increases in imports, and this situation closely mirrored the situation in most other Western apparel manufacturing industries (Jones, 2006). By the mid-1980s, many manufacturers had begun to offshore their manufacturing to low labour cost countries in the East whilst the major sewing machine manufacturers predominantly in the United States, Japan, Europe, and the UK commenced research and development into sewing machinery automation and robotics in an attempt to counteract the impact of cheaper import penetration. Some of the early experiments between the late 1960s and late 1980s focused on developing workerless factories and automated sewing systems using formal and commodity products that used easier to handle fabrics rather fashion products and dimensionally unstable fabric (Jana, 2018). These experiments also included trying to standardise the handling of textile fabrics by freezing or stiffening them, which may have seemed an extreme measure at the time, yet this stiffening approach has more recently been revisited by the Sewbo project in 2018 which has demonstrated the ability to construct the first garment with no human intervention (Sewbo, 2022).

The majority of sewing machines remained as electric-mechanical machines using gears, cams etc. until the advent of circuit boards and microchips which enabled the development of electronic components like the first programmable microprocessor-controlled sewing machines in the 1980s to enable very basic semi-automatic functions such as needle positioning, thread trimming, and programmable stitch count for a determined seam length (Microdynamics, 1981).

Whilst many associated manufacturing industries such as the automotive industry successfully implemented automation and robotics during the period 1980–1995 initially for welding and painting and then assembly (McKinsey, 2018a), the apparel industry lagged behind in automation for two key reasons. First, during the main onset of globalisation at the beginning of the 1980s, the apparel industry opted for the short-term solution of cheaper labour sources in the East because at the time, even though semi-automatic machines had been developed for individual sewing processes to replace some manual labour, there was not a complete automated solution to enable cost-effective sewing production to continue in the West (Tyler, 2009; Jana, 2018; Machova, 2018). Second, both industries work with very different materials, so these comparisons are often generally unhelpful. In the automotive industry, the range of materials do not change often such as steel, aluminium, or fibreglass. The apparel industry has an infinite variability of material and style combinations, the majority of which are subject to constant change with different handling requirements for each. This inability to replicate the human hand for picking up, placing, and feeding different types and different weights of material for the construction of apparel has been and still remains the key barrier to automating apparel manufacturing.

For decades, key developers, such as PFAFF, JUKI, Mitsubishi, Gerber, Lectra, FK Group, Tukatech, Bullmer, and many others, have invested a great deal of money, time, and effort to progress automation for sewing, spreading, and cutting. Headway has been made, however, and current industry challenges make this journey towards all-inclusive automation more compelling and potentially more possible.

5.4   CURRENT INDUSTRY DRIVERS OF AUTOMATION IN THE SEWN PRODUCT INDUSTRY

The first and most pressing driver towards all-inclusive automaton is the need to sustainably manufacture apparel as close as possible to the point of sale to minimise environmental impact by reducing the volume of garments being moved around the world. According to McKinsey (2018a), sustainability is highly likely to be a key purchasing factor for mass-market apparel consumers by 2025. Automation will also be crucial to increasing the financial viability of on-demand nearshoring or on-shoring models (McKinsey, 2018a).

Another key driver supporting the move to automation includes the dramatic improvements in the last ten years in both the quality and the cost of electronic hardware and software needed for the development of automated sewing, cutting, and spreading, making new technologies for automation such as robots much more accessible. However, most small-to-medium size businesses, who now comprise the majority of clothing manufacturers, will need to develop confidence in these new technologies to encourage financial investment to change as investment in new technologies has generally been slow within the apparel manufacturing sector, and currently appears to remain so. A recent survey by Kalypso (2020) involving industry leaders found that 39% of respondents said they would not be investing in technology related to cut and sew within the next two years. Some of the reasons for this included access to talent and skill to support new technology. The justification to implement change was also cited along with vision, funding, and lack of understanding of new technology. Access to knowledge, education, and graduate talent who have this expertise will be key to encouraging investment in automation.

Another urgent driver for automation relates to rising labour costs around the world, as there are currently very few untargeted locations left that can offer a cost-effective labour force. An example that demonstrates the rapid increase in labour costs is China. In 2005, China's labour costs were one-tenth of those in the United States. Just over ten years later, China's labour costs dramatically increased to about one-third of those in the United States (Rozelle et al., 2020). In this short time span, China has developed a fashion industry, which is now the largest fashion market in the world, worth over US$2 billion, and due to its increasing labour costs has increasingly migrated apparel manufacturing to countries such as Vietnam, Cambodia, Bangladesh, and Ethiopia (McKinsey, 2019).

Quick response driven by consumer demand for faster fashion is another key driver for automation. Trends are constantly changing at a rapid pace, as consumers increasingly adopt a see-now-buy-now mentality (Petro, 2018). Therefore, brands need to produce a larger number of small-volume collections with a greater variety of style options that are competitively priced and delivered within a shorter time frame to remain competitive. Achieving these tight schedules in a sustainable manner further strengthens the case for nearshoring and onshoring enabled by new approaches to automation.

Other ways that the consumer is influencing the move towards automation relate to changes in diet, lifestyle, and demographics which have meant that many brands are producing clothes using fit parameters that have become out of date, and this has meant high return rates and elevated costs associated with the administration of returns. In response to this, consumer demand for personalised and made-to-measure apparel has increased, but these new approaches require new sustainable models of manufacture that include nearshoring and onshoring using micromanufacturing, both of which are reliant on high levels of automation (JustStyle, 2020).

The fashion industry needs to respond to these challenges by adopting sustainable and highly automated methods of manufacture. The rest of this chapter will consider the headway that has been made so far to meeting this goal, as well as the challenges that lie ahead for the future.

5.5   SEWING MACHINE AUTOMATION

Sewing machine automation can be classified into three types as follows:

- Simple automates
- Semi-automated machines/automated workstations
- Fully automated sewing machines

5.5.1   SIMPLE AUTOMATES

Simple automates are well documented by a variety of authors, including Hayes and McLoughlin (2008), Tyler (2009), McLoughlin and Mitchell (2013), and Jana (2018), and are the simplest form of semi-automated sewing machines. They are usually of either short or long fixed cycle and cam-controlled, meaning once power is engaged, they must complete the cycle before finishing. They are also not programmable and have limitations in that they can produce only one configuration of sewing such as a buttonhole, buttonsew, bar tack, spot tack, and so on. Fine adjustments such as changing the bight and length of the buttonhole or bar tack are often possible. Although the cam-design enables accuracy and consistency, simple automates are considered as the most basic semi-automatic machines as they are dependent on an operator to not only engage power and start the machine but also both pre and post sewing to pick, place, load, unload, and stack. Opportunities to integrate robots with simple automates will be further discussed later in this chapter.

5.5.2 Semi-automated Machines/Automated Workstations

Some of the first semi-automatic machines developed in the 1980s were made possible by ongoing development in electronic and microchip technology. This assisted the development of semi-automatic sewing machines that utilised Computer Numerical Control (CNC) using stepper motors to transfer minutely accurate movements to mechanical parts, delivering a high degree of precision for sewing in the X and Y directions with attachments such as sewing clamps or jigs (Hayes and McLoughlin, 2008).

Over the last 40 years, ongoing development in semi-automatic machines for both woven and knitted materials has meant that individual semi-automatic sewing machine now exists for most individual sewing operations. What is still limiting the movement towards full garment automation is the ability to join all the individual operations together to make a garment. Some examples of semi-automated sewing operations include:

- Patch and cargo pocket setting and hemming
- Welt and Jett pockets
- Sequential buttonhole and buttonsew
- Profile auto-jig for small components such as collar, cuff flap
- Seaming and hemming for components such as back yoke, side seam, and shoulder seam
- Cuff attach
- Darts and pleats
- Belt loop setters
- J stich on trousers
- Continuous zip tape application
- Sleeve setting

Most semi-automatic machines are also commonly referred to as workstations because of their size, and most still require varying degrees of human interaction such as loading and unloading despite continuous developments in work aids designed to further reduce human interaction in the overall sewing process. The variety of work aids that have been used to support automation include compressed air, which is predominantly used to move machine parts such as the presser foot. Other work aids include edge guides, seam and hem folders, and slack-feeders. In more recent years, the types of work aids that were traditionally housed within a microprocessor control panel including thread trimming, raising and lowering the needle bar, back tacking, and stitch counting have been replaced by a reprogrammable digital control panel that is now built into most semi-automatic machine as a standard practice.

Stackers have also been an important work aid in supporting steps towards fully automated sewing as they enable the disposal and stacking of sewn components in preparation for the next cycle. There are a variety of stackers that accommodate large and small component stacking, but in essence these work aids are designed to replace human labour in the movement and stacking of sewn components. Further details of the variety of stackers currently in use in the apparel industry are comprehensively

FIGURE 5.1   Juki APW 896N lockstitch automatic welt pocket workstation.

Source: Courtesy of www.Juki.com

outlined by Jana (2018). In terms of loading and placing, developments in pneumatics have enabled the automatic loading of components that need to be attached such as zip guards, pockets, pocket flaps, and so on. Stacking, picking, and loading systems all have the potential to be replaced by robotics and will be discussed further later in this chapter.

All of these developments have enabled the semi-automatic pre-assembly of individual sewing operations; however, assembling them together to make a complete garment is the next challenge on the road to automation, and some of the latest steps that have been taken along that road will be outlined later in this chapter.

5.5.3   FULLY AUTOMATED SEWING SYSTEMS (AUTOMATED TRANSFER LINES)

Attempts to create fully automated sewing systems or transfer lines that can automatically construct a complete garment can be traced back to the 1960s, when key machine developers PFAFF recognised the importance of automating sewing production. This resulted in the Transfer Street System, which automated all of the processes on the left front of a man's shirt, including front strap construction, buttonholing, pocket attaching, and hemming. The system used a conveyor system to transport components from a cut stack to sewing heads that used stitching clamps and jigs to guide sewing and pressure-sensitive tape for pick-up. According to Tyler

(2009, p. 187), 'pre-folded and pressed pockets were fed from a magazine and transferred in a clamp device to the machine for attaching pockets'.

Although it had many limitations and was at this time commercially unviable, many of its features assisted to pave the way for the development of the wide variety of semi-automated workstations that were developed from the 1980s onwards (Tyler, 2009; Mcloughlin and Mitchell, 2013).

Further attempts to develop a transfer line to automate apparel construction were undertaken by the American Textile Clothing Technology Corporation (TC2) and Singer in the 1980s. According to Tyler (2009, p. 190), the sections of the line consisted of:

- an automatic loader to insert parts to be loaded into the transfer line;
- a viewing line that allowed the automatic vision system to recognise the parts;
- a robot that can fold and align the edges;
- a transfer door that slides the parts to the sewing station;
- a seeing unit with feed belts and a sewing machine under computer control.

Individual prototype lines were developed for the assembly of different components such as jacket sleeves, coat backs, or trouser legs, but none of them proved to be commercially viable.

Throughout the 1980s, the Japanese Ministry of Trade and Industry pumped significant sums of money into their Technology Research Association for research into the development of automatic production systems for clothing, some of which utilised robotic systems for sewing operations; however, according to Tyler (2009), the technology became extremely complex and therefore economically unviable.

Some of these early advancements can perhaps be seen as the beginnings of the use of robotics to support the successful automation of clothing manufacture.

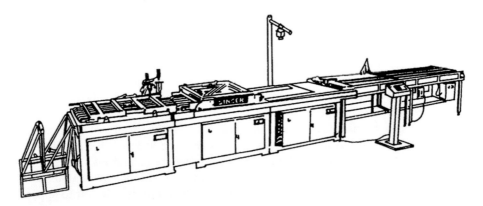

FIGURE 5.2   TC2/Singer's prototype coat back unit.

Source: Courtesy of Tyler (1989)

5.6   ROBOTICS IN AUTOMATED SEWING MANUFACTURE

5.6.1   ROBOT MORPHOLOGY

According to the International Federation of Robotics IFR (2022), industrial robots are programmable task-performing machines that can automatically carry out a complex series of tasks or actions with speed and precision that would otherwise be undertaken using human labour.

The first industrial robot was developed by George Devol and Joseph Engelberger in 1959 and weighed two tonnes. It was controlled by a program on a magnetic drum and used hydraulic actuators with an accuracy of 1/10,000th of an inch. Industrial robots continued to be developed throughout the 1960s and 1970s and were successfully integrated into industries such as automotive and aerospace to reduce labour costs and improve efficiencies (IFR, 2022).

Traditional industrial robots had many advantages but also several limitations which included safety and due to their size, strength, and unpredictability worked separately from humans behind caged or fenced locations. Their application in small-to-medium size enterprises (SMEs) was a further challenge due to the floor space requirement to house them.

There are many different types of robots which according to Intel (2022) can be divided into mobile and stationary robots. Mobile robots such as Autonomous Mobile Robots (AMRs) find applications in the health sector to deliver medication or disinfect surfaces. Their mobility has relevance to apparel manufacturing and could be used, for example, to deliver cut components from a cutting room to a semi-automated workstation. Automated Guided Vehicles (AGVs) could also be potentially used in apparel manufacturing to deliver materials and move items such as heavy fabric rolls in factories. There are also humanoids that perform human-centric functions such as a concierge service and have human-like forms, perhaps these may replace some aspects of operative-based sewing in the future.

In terms of stationary robots, these include cobots and articulated robots which can have up to ten rotary joints for greater degree of motion. Just over a decade ago in 2008, a Danish company called Universal Robots launched the UR5, which was the world's first collaborative robot capable of sharing the same workspace with people as they were developed with safety sensors and cameras so they could work safely with humans as work assistants and for this reason are called cobots (Universal Robots, 2022). The UR5 was the first cost-efficient and user-friendly cobot that was designed to meet the needs of the small-to-medium-size manufacturing community that had largely been excluded from the traditional robot market which was previously too costly and complex.

Cobots can be used as a work assistant for a wide variety of applications from manufacturing to health and even personal care at home for people with disabilities. They are often applied in situations that are dangerous such as overhead work, or for monotonous and repetitive tasks, or highly meticulous tasks where zero defects are required, and this makes them highly relevant to apparel manufacturing.

A cobot is relatively simple in terms of its make-up, and the key constituent is the main body which is also referred to as the robot which is manufactured by several brands including Doosan, FANUC, Universal Robots, Epson, and Kinova. All cobots need to be fitted with an interchangeable end-of-arm tool that suits the operation to be performed. There are a wide variety of end-of-arm tools such as two- and three-finger gripper tools, magnetic grippers, vacuum or pneumatic grippers, even dispensing cartridges with syringes for joining or gluing, and so on. Safety tools such as sensors or cameras are an essential ingredient in a cobot built to enable safe working conditions for co-working.

Beyond these three core items is the requirements of a table and robot mount to house the robot, plus cables, pneumatic connectors, and software to enable the cobot to run or learn a new process. Most cobots have been designed to be easy to set up and use, and most are built on a plug-and-play basis.

Cobots can be easily programmed or taught to learn from humans, whereby just one demonstration of a specific process/movement is needed, such as loading a sewing machine with a component. Cobots can record and then replicate this movement/process as many times as needed allowing the human to undertake other work.

FIGURE 5.3a Robot arm.

Source: Courtesy of www.universalrobots.com

FIGURE 5.3b   Gripper robot end effector.

Source: Courtesy of www.piab.com

FIGURE 5.3c   Robot safety sensor.

Source: Courtesy of www.sick.de

Cobots such as the KUKA LBA iiwa can be considered as a second wave of robotic technology that has been successfully integrated into the automotive industry to replace smaller tasks previously undertaken by humans that are considered dirty and monotonous such as applying adhesive sealants. Cobots can accurately execute the latter motion sequence independently, much quicker and on a continuous and highly accurate basis without ever stopping and are therefore vastly more efficient than a human (KUKA, 2022).

In the last ten years, the cobot has been introduced to many other manufacturing sectors but has only begun to infiltrate the sewn product industries in the last couple of years in a superficial way via a variety of research and development projects which will be discussed in further detail later in the chapter.

It is surprising that uptake has so far not been faster and more pronounced within the apparel manufacturing sector, especially in terms of the cobot's ability to complete highly monotonous, meticulous, and even dirty tasks in an industry that is needing to replace labour-intensive manufacturing processes.

5.6.2 Current Drivers and Enablers of Advanced Automation with Robotics in the Manufacture of Sewn Products

Manufacturers across a variety of sectors are phasing out human labour and implementing robots to enable their manufacturing activities and material flows to be fully automatic. Examples include FANUC robots that are being used for intelligent waste management; Stäubli six-axis robots have been introduced by the Leupolz Emmental dairy in Germany for handling giant cheesecakes; and the Bhe Belgian BESIX group has introduced a KR QUANTEC robot from KUKA for 3D printing of concrete (IFR, 2022).

A key driver for replacing humans in the industrial manufacturing sector is the ability to liberate humans from manufacturing jobs that are often dangerous, such as working with chemicals or biohazards and welding, or simply from the monotony of repetitive manufacturing work. Other reasons that are driving the switch from human to robot are based on the need to achieve sustainable manufacturing by reducing waste by eliminating errors, thus improving quality which in turn improves consistency and profitability. But another more pressing driver is the need to reduce reliance on human labour due to rising labour costs, which have continued to rise since the 1990s (McKinsey, 2018a).

The benefits of using robots to automate manufacture have been evident since the 1980s, when the automotive industry switched humans for robots for undertaking dangerous work involving welding and paint spraying. The electronics industry also realised the wide-ranging benefits of robots in the 1990s followed by Amazon for warehousing in the millennium (McKinsey, 2018a). Key enablers for this technology currently include ongoing reductions in the price of robots which Tilley (2017) argues have steadily reduced by over 50% in the last 30 years in comparison to labour costs.

In terms of the skill base needed to enable and maintain the use of robotic technology in manufacturing, according to Tilley (2017), people with the skills required to design, install, operate, and maintain robotic production systems are becoming more widely available, and advances in technology and software have made the assembly and installation and maintenance of robotics faster and less costly than before.

This is certainly true for sectors such as automotive, electrical, and warehousing and even automated cutting technologies in the apparel industry where research into the use of robots and new gripping technologies for the picking and stacking of cut work is being undertaken by companies such as Bullmer (Germany). But this is not quite the same scenario for the sewing sector due to the ongoing complexities and limitations of automating the sewing of complete garments.

It is however possible to integrate cobots for the sewing of individual components, and this has been more recently evident within the sewing technologies for the

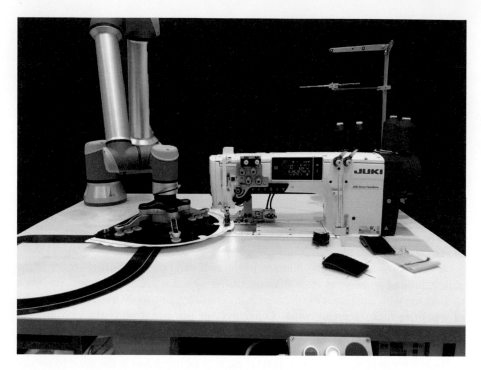

FIGURE 5.4   Universal Cobot robot feeding cut work to a Juki sewing machine.

Source: Author's own

automotive interiors and airbag industry whereby CNC profilers and jigs guided by robots are being used to automate their construction.

Robots are also being used to 3D stitch car seats and dashboards, but these developments have so far failed to filter across to the apparel industry because at this point in time, most manufacturers believe labour is still cheaper and less risky than technological investment in automation. This point is supported by a recent survey by Kalypso (2020) involving industry leaders in the apparel industry which found that 39% of respondents said they would not be investing in technology related to cut and sew within the next two years with additional reasons cited to include access to talent and skill to support new technology, justification to implement change, vision, funding, and lack of understanding of new technology. All of these reasons have meant that the apparel manufacturing sector remains locked into using traditional human labour for manufacturing, and this cannot continue for much longer.

Companies such as Vention might offer the apparel industry a way forward with the integration of robotics. Vention is a recent start-up company that has emerged in the last six years. Its mission is based on the democratisation of robotics technology; it has developed an easy-to-use 3D design tool called 'Machine Builder' that customers can access to design and build their own robotics-driven manufacturing floors and production lines. Drop-down menus in their specialist software are loaded with a comprehensive catalogue of all the components that will be needed for each design/build. All components that are offered on its site are sourced from leading robotics companies, such as Doosan, FANUC, Universal Robots, Epson, and Kinova. Components include structural framing, panels and tops, robot mountings, robot end-of-arm tools, as well as a variety of pneumatics, cabling, controls, and motors, all of which are pre-tested to ensure they can be easily joined together 'Ikea-style' using a simple plug-and-play approach with no drilling or cutting. A further key enabler of this approach is that commissioning is straightforward as there is no complex assembly, no configuration, and no programming prior to commission. Vention's 3D Machine Builder also provides a realistic yet simple design and simulation environment so users can evaluate their automated system's performance before making any upfront investment. Such an environment not only verifies that the automated process will work, but it also confirms the estimated cycle time and total investment. With this information, users of the platform gain insight into their expected ROI before purchase. To support further improvements in robotics technology, Vention has also developed a community forum for the sharing of best practice and advice.

Another key driver that is assisting to further enhance advanced automation with robotics in the manufacture of sewn products is AI. According to Tilley (2017), AI is already being used in various manufacturing sectors to complete complex tasks such as 'altering the force used to assemble two parts based on the dimensional difference between them, or by selecting and combining different size components to achieve the desired final dimension' (Tilley, 2017, p. 8). Both examples are highly relevant to the complex tasks associated with garment sewing and construction, and there are other useful examples operating in the automotive industry that provide a vignette of what automated apparel manufacturing might look like in the future. Tilley (2017) describes a highly sophisticated automotive manufacturing system that can balance the whole production lines using AI by self-adjusting the speed of sections of the line. It is also able to immediately switch to a totally different product without the need to change programs or retool due to interconnected digital machinery and equipment. Both examples are highly desirable and essential for demand-driven micromanufacturing of sewn apparel, yet are currently unattainable with current human-centric manufacturing approaches. A realistic future for automation in the apparel industry should consider lessons learnt from other associated manufacturing sectors that have carefully and successfully balanced the introduction of consecutive levels of automation over a lengthy period of time such as the automotive industry. Other recent examples include the Adidas SPEEDFACTORY, established to experiment with new manufacturing models by substituting expensive human labour for robotic manufacturing in Germany. The factories lasted less than three years and had to close partly because the automation they implemented had limited applications for the diverse range of products that the company make. Therefore, as argued by Tilley (2017), decisions on automation must be holistic and systematic and align with current and future needs of the business.

An area of apparel manufacturing that has successfully achieved sophisticated levels of automation is the fabric spreading and cutting of sewn products which will be discussed in the next section.

5.7   AUTOMATION FOR FABRIC SPREADING AND CUTTING

5.7.1   AUTOMATED FABRIC SPREADING AND CUTTING
SUPPORTING INDUSTRY CHALLENGES

The first automatic cutting machine was the Gerber Cutter S-70, developed for industrial automated cutting of fabric by Gerber Technologies in the late 1960s. It is still regarded as one of the most significant and successful inventions in clothing manufacturing-automation of the twentieth century. Part of the reason for its success is based on the fact that early models were proven to be extremely successful and immediately gained recognition and acceptance because they actually did what the advertising said they would do in terms of their ability to accurately cut virtually any fabric with just a simple change of the cutting knife or blade type.

Another reason for continued user acceptance of this technology is based on the fact that manual cutting and spreading have well-known limitations, which include being time-consuming and therefore expensive, and human error in the form of cutting faults can also be irreversibly costly. Safety is another key aspect, as manual cutting equipment can be extremely dangerous unlike automatic cutting machines where cutting tools are fitted with highly secure safety guards.

The key driver that inspired the development of this important innovation back then remains just as relevant today, which is based on the need to reduce the apparel manufacturing industry's reliance on cheap human labour by adopting advanced automated technologies and in doing so, assist to reduce operating costs, achieve greater efficiencies through improved quality by eliminating human error leading to less wastage and more sustainable approaches to manufacture.

Over the last 50 years there have been ongoing improvements and refinements in automated spreading and cutting technologies by companies such as Lectra/Gerber (France), Assyst Bullmer (Germany), Morgan Technica (Italy), FK Group (Italy), Shima Seki (Japan), Tukatech (United States), and many others. Some of these refinements have only recently been enabled by ongoing improvements in both the quality and the cost of electronic hardware and software, and this has led to research and development to further automate, for example, an aspect of cutting that has largely remained manual, namely the picking and stacking of cut work using cobots.

Other recent advancements include the development of an augmented reality camera system by the FK Group (Italy) that enables the highly accurate and real-time planning and cutting of checks and stripes even on multiple plies using special optical cameras plus dedicated software (FK Group, 2022).

These ongoing developments in spreading and cutting have all assisted in meeting new and emerging industry challenges which have been outlined earlier in the chapter but for convenience are summarised as follows. First, *sustainable manufacturing* is supported through the automation of fabric spreading and cutting technologies because automation reduces human error and this in turn reduces fabric waste. This is particularly relevant to cutting technologies where mistakes such as over cuts can be devastating when bearing in mind that up to two-thirds of the cost of a garment is the cost of fabric. Automated spreading can also assist in reducing fabric wastage, for example, as the spreader can accurately locate and cut ply ends consistently.

Second, the substitution of human labour with automated spreading and cutting technology increases the financial viability of nearshoring and onshoring using new models of automated manufacture such as micromanufacturing. It is these new models of manufacture that can also meet consumer demand for smaller and quicker collections and indeed the trend towards personalised and made-to-measure apparel. For example, a bespoke fully tailored jacket can take several hours to cut using manual methods but only a few minutes to cut using an automated single-ply cutter, meaning there is a huge overall saving of labour costs and handling. Another example is an item of intimate wear such as a bra that involves extremely small measurement tolerances requiring a high degree of accuracy. The automatic cutter can quickly and effortlessly meet this level of precision.

The rest of this section will provide a brief summary of the advantages and limitations of automated spreading and cutting versus manual approaches to spreading and cutting, which are still heavily used across the global apparel industry.

5.7.2   ADVANTAGES AND LIMITATIONS OF MANUAL AND AUTOMATIC SPREADING

Prior to the introduction of automated cutting and spreading machines, spreading was a manual operation requiring two people, one at either side of the table to unreel and lay cloth. The process of manual spreading is time- and labour-intensive, and the tensioning of each ply can be irregular as determined by the skill of the operative. This results in fabric lays needing to remain on the cutting table for up to 24 hours to relax in preparation for cutting. In addition to this, manual ply end cutting can accumulate fabric waste, and operator fatigue is likely due to the continuous handling and motion involved. However, the use of spreading with a carousel delivers a number of advantages. First, it is quicker than manual laying, and a major advantage of the carousel is the greater consistency of control over fabric tension, so time for fabric relaxing can be reduced. Most carousels are mounted with automatic end cutters such as a simple circular knife mounted on track, and this along with end catchers and clamps can be used to ensure ply ends are of same length which can assist in reducing fabric wastage.

In comparison to manual spreading or spreading with a carousel, semi-automated and fully automated spreading brings a wealth of benefits, for example automatic spreaders can spread up to 100 metres per minute in lay heights that can exceed 22 cm. They can handle large and heavy fabric rolls that can weigh in excess of 150 kg and have computer-enabled tension control, which minimises the need to relax fabric, thus eliminating table-blocking and enabling continuous spreading of new lays. Operator fatigue is also minimised as many semi-automatic spreaders have travel-on-platforms to enable the operative to observe the laying process and to check and remedy faults within the fabric or on the selvedge. Fully automatic spreaders are capable of automatically detecting and removing fabric faults during the laying-up process.

5.7.3   ADVANTAGES AND LIMITATIONS OF MANUAL AND AUTOMATIC CUTTING

In the absence of automated cutting machinery, hand cutting using shears or electric knives that can include straight knives, band knives, or circular saw is used to cut single or multiple plies.

The key advantage of using the electric straight knife, as opposed to hand shears, is that it can be used for intricate jobs that need long clean cuts such as small radius cuts and intricate patterns. The straight knife allows for tight turns and minimal distortion between top and bottom plies and cutting height up to 29 cm. The straight knife is equipped with a one-touch automatic sharpener and dual speed to prevent fusing when cutting fabrics with a high thermoplastic content. The electric straight knife can also be used for notching; however, it requires a lot of skill and training. There is also an enhanced safety risk of amputation if not using chain mail gloves. Overall, the speed of cutting depends on the skill of operator, and expensive irreversible mistakes are possible.

Unlike the electric straight knife which is manoeuvred through a lay by the operator, the band knife is stationary, and the lay is moved to the blade on the band knife. It can be used for extremely intricate or accurate cutting where both hands are required to move fabric around the blade, and this is not possible with hand-held straight knife. The band knife can also be used for notching and requires a lot of skill and training. Just like the straight knife there is a high risk of amputation and other serious safety risk if not using chain mail gloves. Likewise, the speed of cutting depends on the skill of operator, and expensive irreversible mistakes are possible.

Circular saws have a disk-shaped blade and are only appropriate for cutting straight lines or very large radius curves. Intricate shapes or tight curves are not possible, but high ply depth is possible when cutting straight lines and large radius curves only.

Single-ply cutters are useful for prototype development, or bespoke made-to-measure garments, and particularly relevant for styles that have a high volume of components such as tailored blazers or waterproof performance coats. They are highly accurate, specifically for intricate designs or bias cutting. Automatic cutters and spreaders require specialised software that support digitised patterns to be imported or for digital patterns to be created, so that highly efficient digital markers can be generated. Detailed information concerning the latter is contained within a separate chapter of this book that focuses on digital pattern cutting systems.

Other advantages of single-ply cutters are that most can cut more than one ply depending on fabric thickness, and this provides further savings in labour, time, and overall costs. So, although the initial set-up costs are high, the payback can be quick especially for busy sample rooms or a small business where access to and affordability of additional labour may be limited.

Automatic sew bar stickers that support component identification can also be enabled on single-ply cutters that are equipped with a conveyor-style movable bed. This enables each component to be labelled with a batch number which supports component identification when cut work is bundled. This prevents garment with mixed-size components being bundled together.

Multi-ply automatic cutters work on the same principle as the single-ply cutter. You can achieve a high degree of accuracy so long as the initial pattern input is correct. Bulk production can be cut in fraction of time, and automatic cutting eliminates traditional health and safety issues linked to operator handling, and there is no need for additional cutting equipment such as band knifes for very detailed work. Just one operative will be needed if equipped with a self-loader and spreader.

Laser cutting, as the name implies, works by directing the output of a high-power CO_2 laser to the material to be cut. The material then melts, burns, vapourises away, or is blown away leaving an edge with a high-quality surface finish. Most industrial laser cutters are large single-ply machines. The advantages of laser cutting over mechanical methods are similar to single-ply cutting as there is no physical contact with product during cutting, meaning there won't be any slips or mistakes and no contamination such as oil or finger nicks. Unlike knives and blades that need to be replaced on manual or automatic cutting equipment, there is no wear on a laser and no parts to replace. Laser cutters enable extremely intricate cutting and depending on the fabric composition can also be set to heat seal the fabric edge to prevent fraying. Some fabrics do not cut well such as wool and leather and other natural-origin textiles which can result in charred edges and acrid fumes, for example wool will smell like burning protein. Some materials cannot be cut using lasers such as PVC because it gives off highly dangerous gases such as hydrogen chloride, which is toxic to humans and highly corrosive to laser equipment. Lasers can also be used to etch the surface of fabric such as denim as a sustainable approach to simulating authentic denim finishes traditionally achieved through wet processing.

A water jet cutter is a tool capable of slicing into materials using a jet of water at high velocity and pressure. Water jet cutting is commonly used for cutting polymer-based textiles but also for leather cutting because no fumes are given off with this method, and there are no charred edges or acrid fumes. Textiles which readily absorb water are not so suitable. Water jet cutting is considered a green technology, as there is no hazardous waste, and it uses very little water (a half gallon to approximately one gallon per minute depending on the cutting head orifice size). Used water can be recycled using a closed-looped system, and water waste is usually clean enough to filter and dispose of down a drain.

Dye cutting, similar to the principle of a cookie cutter, can accurately cut intricate designs and can be used for any type of material but is particularly relevant to staple products as the major limitation of this cutting method is that new dyes need to be made for every style change.

From this brief summary, it is clear to see the advanced benefits of automated spreading and cutting in terms of speed, accuracy, and safety along with other efficiencies such as cost, waste and labour reduction, and quality enhancement. There are currently few other examples of automation in the apparel manufacturing industry that so fully meet these wide-ranging benefits.

5.8 RECENT ADVANCES AND INNOVATIONS IN MANUFACTURING AUTOMATION FOR LAYING, CUTTING, AND SEWING

This section will provide a summary of some of the most recent advances and innovations in automation for apparel manufacture. For ease of reading, the section has been divided into the following subsections:

- Advances in fully automated garment construction and 3D sewing
- Advances in automated and semi-automated 2D sewing
- Advances in spreading and cutting technology
- Advances in seam bonding and welding

5.8.1 Advances in Fully Automated Garment Construction and 3D Sewing

In 2016, the Sewbo project demonstrated the ability to construct a garment with no human intervention (Sewbo, 2022). The principle behind this development involved stiffening the fabric components that would make up the top with a water-soluble PVA emulsion which when dry created stiff fabric. The stiffness of the fabric enabled an industrial robotic arm to lift the components and feed them into the sewing head of a domestic sewing machine. After all the seams had been joined, the garment was rinsed in water to eliminate the stiffening agent. As outlined by Gries and Lutz (2018), some critiques have argued that the washing process may adversely impact the cost-effectiveness of the overall process; this has not yet been validated but perhaps there is potential for this within the casual wear and denim sector where wet processing of apparel is standard practice.

5.8.1.1 SoftWear Automation: Sewbots

SoftWear Automation is a highly specialist manufacturing company that makes fabric-based products for the home goods sector as well as on-demand and custom apparel items such as basic T-shirts for the fashion industry using fully automated handling and sewing technology that involves an advanced vision and robotics system that has taken several years to develop, supported by funding from sources such as DARPA, The Walmart Foundation, and the Georgia Research Alliance. According to Fraser (2017), the Sewbot system can produce nearly twice the volume of T-shirts in the same time period as a human sewing line of ten workers.

The handling of cut components is enabled by a combination of a conveyor system and a robotic arm called Lowry. The conveyor is a specialised ball-style conveyor

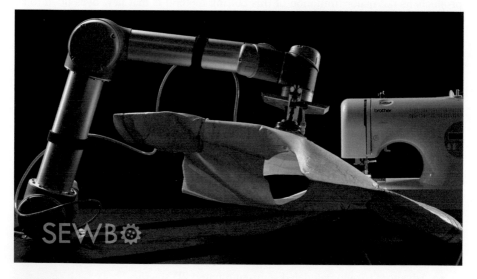

FIGURE 5.5 The first garment constructed with no human interaction.

Source: Courtesy of Sewbo Inc. Photography by Zack DeZon

system, which enables components to be transported in any direction unlike most linear conveyors that can travel only forwards and backwards.

The robotic arm is fitted with specialised end-of-arm vacuum grippers that can lift and place fabric components for sewing. The sewing process is supported by a computer vision system which maps the surface of the material as it sews and has high levels of accuracy of up to 0.5 mm.

5.8.1.2   Adidas SPEEDFACTORY: Fully Automated Sports Shoe Manufacture

Two Adidas SPEEDFACTORY was established in 2017 to experiment with new manufacturing models by substituting expensive human labour for robotic manufacturing in Germany and in the United States. The operating principle was to reshore manufacturing to facilitate on-demand manufacturing thus supporting sustainable design and manufacture whilst reducing reliance on human labour but implementing innovative labour-saving 3D printing, cobots, and computerised knitting to make highly innovative sports shoes. The factories proved to be initially successful and operated for nearly three years. The decision to close them was based on lack of flexibility of the technology which had limited applications for the diverse range of products that the company now make (Speer, 2019; The Robot Report, 2019). This example has been included in this section because this advancement has provided essential learning for future manufacturing automation projects.

5.8.1.3   PFAFF (Germany) KSL Robot 3D Sewing Units

3D sewing has been traditionally used within the construction of composite materials since the 1980s for industrial applications particularly in aerospace. Research and development in the late 1990s by Moll (1997) indicated its potential use for other sectors

FIGURE 5.6   SoftWear Automation's sewbot system.

Source: Courtesy of www.softwearautomation.com

including apparel manufacturing, yet this technology has remained generally underdeveloped for this particular end use (Song et al., 2022). However, recent developments by PFAFF have showcased its use for the 3D stitching of automotive interiors such as dashboards using a sewing machine mounted on a robotic arm. Unlike the traditional sewing approaches where the material is presented to the sewing machine, with 3D sewing, the sewing machine is presented to the material which is draped over a specially shaped form/stand. The programmable robot enables the sewing machine to precisely follow the sewing pattern. This is a useful advancement which represents a refreshed step forward to solving the issue of handling limp material for garment construction by applying technologies that avoid handling fabric and instead handling machinery.

5.8.2 Advances in Automated and Semi-automated 2D Sewing

The PFAFF KL311 is a programmable CNC sewing unit with a 360° orbital sewing head that can rotate and sew in any direction, meaning the stitch formation will always be running in the same direction, which has always been a limitation of CNC machines that run in the X–Y direction. It has a large sewing field (2000 mm × 1000 mm) and is fitted with vision camera technology, which allows it to automatically reposition stitching during sewing. This is especially important for decorative stitching that has been designed, for example, to topstitch breathability holes on the back of a car seat.

5.8.2.1 JUKI MH-2960V/Robot Arm for Feeding Cut Components Held in a Sewing Clamp

JUKI has developed a prototype sewing system (Figure 5.4) with an integrated robotic arm from Universal Robots that can feed a cut component to a twin needle chainstitch machine. The fabric component is fixed to a sandwich clamp, the clamp is made from smooth plastic so the suction head on the robot arm can grip the clamp. The robotic arm can then rotate the clamp in a circular motion. The clamp uses magnets to hold the material in place and ensure no damage to the material (as clamps can damage the airbag fabrics).

5.8.2.2 JUKI AW-3S Automatic Bobbin Thread Winding and Feeding Device

JUKI has developed the AW-3S automatic bobbin thread winding and feeding device which can automatically calculate the amount of bobbin thread required for a particular sewing pattern and change the sewing thread automatically. Both features enable greater efficiencies and endless sewing as the thread never runs out but also assists in reducing operator fatigue from having to undertake these processes manually.

5.8.2.3 PFAFF 3834 Sleeve Setter with Automatic Thread Tension Adjustment for Different Ply Heights

The machine enables the setting of the sleeve through programming on the digital control panel so that you can input different degrees of feeding in the circumference. The key advancement is the use of the roller feed and an inbuilt sensor that can detect different ply thicknesses and adjust the needle thread tension to compensate for the different ply heights.

5.8.2.4 PFAFF 3688 Automatic Pocket Setter for Stretch Fabrics

Automatic pocket setting machines are not new; however, advancements in stretch fabrics have necessitated refinements in pocket setting technology. The refinements are based around having stronger transport motors (servo drives) to optimise the timing of transport feeding. The result is the ability to reduce the thread tension to enable less pucker when, for example, stitching stretch fabrics.

5.8.2.5 PFAFF 3590-VARIO: Programmable Large Area Sewing Unit

This is a specially designed programmable large area sewing unit for sewing sports shoes, automotive interiors, technical textiles, and workwear. Sewing programmes and software changes are entered via an SD card, and existing CAD data sets can be used with the machine's programming system (PSP) to create sewing and cutting programmes.

The machine is designed with a vertical hook, enabling multi-directional sewing, and has a height-adjustable floating presser foot suitable for use with work clamps, which have been designed in a way that enable the development of sewing jigs by local technicians.

5.8.2.6 Vi.BeE.Mac: Model 3022 BHE: 2-in-1 Lockstitch and Chainstitch Machine

This is a unique two-in-one machine that can be quickly and easily converted from a 301 lockstitch to a 401 chainstitch. Trained mechanics can convert the machine in less than 20 minutes. It is not a new development but has new relevance as it is an incredibly flexible machine that accommodates style changes without the need to buy another machine and therefore has relevance to new manufacturing models that can support on-demand and personalised apparel.

5.8.2.7 Vi.Be.Mac V900.1 Fully Automatic Keyhole Buttonholer for Trouser Fly

The V900.1 is a fully automatic keyhole buttonhole machine with loader and unloader. The distance between the buttonholes is programmable with a touchscreen panel. A stack feeder automatically lifts and feeds front fly components using compressed air suction to a linear clamp that transports it to the sewing head which then inserts the desired number of keyhole buttonholes required. It then stacks the components ready to be transported to the next operation. The loader unit works with a patented device which does not damage the fabrics during pick up and place. This machine can be linked to a manufacturing execution system that can control maintenance and production management.

5.8.3 Advances in Spreading and Cutting Technology

Bullmer GmbH, located in Stuttgart Germany, has developed an automatic cutting system that is equipped with a cobot from Universal Robots that is used to sort and stack cut components. The robotic head can be equipped with a gripping system that is matched to the material. The robot is controlled by a separate software application that can communicate with the primary cutting software. The initial trials that

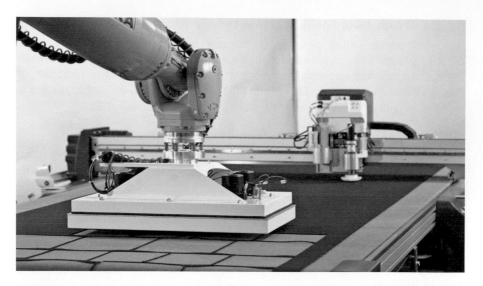

FIGURE 5.7   Robotics integration with Bullmer cutting machine.

Source: Courtesy of www.Bullmer.de

Bullmer has been undertaking originally commenced with a Kuka robot equipped with a special suction head which worked well with heavy textiles such as carpets and jeans but is not suitable for lighter weight materials used for apparel such as cotton which are air permeable. This led to experimenting with a variety of different end-of-arm tool manipulators such as claw style finger grippers, needle grippers, or the glue-based grippers.

5.8.3.1   FK Group Augmented Reality System for Plaid, Check, or Stripe Matching

Other recent advancements include the development of an augmented reality camera system by the FK Group (Italy) that enables the highly accurate and real-time planning and cutting of checks and stripes on single or multiple plies using special optical cameras driven by dedicated software. This advancement is particularly relevant to the development of personalised apparel where luxury bespoke materials and textiles are individually created as one-off pieces of cloth, the operator can complete a simulation of a plaid, stripe, or check match. AI can also be used for bulk production using algorithms to create intelligent lay plans and high-utilisation lays with minimal or zero waste.

5.8.4   ADVANCES IN SEAM BONDING AND WELDING

5.8.4.1   Vetron Typical Europe GmbH (Kaiserslautern) Cold Gluing Robot

The Cold Gluing Robot has been developed to seal seams using a new cold glue technique that is applied by a robot. This process has been developed for automotive interiors but can also be applied to outdoor apparel. This innovation is based

FIGURE 5.8   Vetron typical cold gluing robot.

Source: Courtesy of Vetron Typical cold gluing robot—Google Search

on the ability to seal seams with adhesive tape on the reverse side using a robot to apply the adhesive tape in an automated process which also cuts it off automatically at the end of the gluing cycle using a tape cutter.

5.9   FUTURE OF AUTOMATION AND ROBOTICS IN THE APPAREL MANUFACTURING INDUSTRY

The 60-year journey to automate garment manufacturing has been a slow and arduous one. During this period, the cutting and spreading sector has been the most successful with fully automated commercial cutting and spreading solutions that have widespread and successful implementation.

However, the success for sewing and handling has been less fruitful, hindered for two inter-related reasons. First, unlike related industries such as the highly automated automotive industry that have standard materials, the sewing industry has an infinite array of materials that are constantly changing with new developments all the time. Consequently, sewing machine developers have struggled to find effective technologies that can replicate the sensitivity of the physical human handling involved in the construction of garments involving an infinite array of diverse materials. Second, the sewing industry's willingness to invest in automation has generally been limited as there has always been the alternative of outsourcing the labour-intensive work of handling and sewing to cheaper labour sources in the East. More recently, in the last five years, manufacturers have realised that the supply of cheap labour sources is running dry, and the additional pressure from consumers to meet sustainable manufacture through measures such as nearshoring and onshoring, coupled with additional industry challenges such as the recent global pandemic, on-demand manufacture, plus the positive developments in computer hardware, software, AI and robots have given rise to a refreshed drive towards automation. This has culminated in a dramatic leap forward in apparel manufacturing automation by companies such as SoftWare Automation and Adidas who have remarkably defied all odds and achieved the onshoring of bulk-produced staple garments such as T-shirts (onshored to the United States) and knitted sports shoes (onshored to Germany), enabled by fully automated manufacturing methods, and are continuing to interrogate lessons learnt to enhance future developments. The learning from both examples provides an excellent blueprint to the wider industry for further research and development.

The automation of non-staple on-demand apparel is still a challenge because the sophistication of current manufacturing automation is not yet able to manage the complex designs or extreme variability of fabrics. It is however currently possible for individual garment components to be semi or fully automated using semi-automatic or programmable workstations that use work clamps with sewing jigs which have been more recently further automated by replacing the human labour needed to feed or control the jig with a robotic arm or cobot. The opportunity to implement robots in this way has been due to the progressive reduction in the cost of robots, as well as their widespread availability and their ease of use which ten years ago would not have been possible.

Despite successful research attempts by Sobo to automate garment construction by modifying the fabric using stiffening agents to facilitate robotic fabric handling, it is still not yet possible to fully automate the construction of garments that are more complex than a T-shirt. However, ongoing developments in robots and AI-assisted vision systems will certainly support the future progression of automation. In 2017, the *New Scientist* reported that there was a 50% chance that robots would outperform humans in all tasks within 45 years (Revell, 2017); this is certainly the case for cutting and spreading where research continues to find other applications for robots in this sector such as the use of robots for picking, stacking, and transporting cut work to sewing stations using Autonomous Mobile Robots (AMRs) or Automated Guided Vehicles (AGVs).

Furthermore, bearing in mind the first commercial cobot, the UR5 was only introduced by Universal Robots in 2008, new scientific developments in robot technology are increasing rapidly. For example, in 2022, researchers at Imperial College London have designed a malleable robotic arm that can be guided into shape by a person using augmented reality (AR) goggles (Brogan, 2022). Unlike most current robot arms that have rigid limbs and firm joints, this malleable third arm represents a further development of the cobot and might act as an extra limb to further share workloads.

But the future of automated apparel manufacturing will not be solved solely by new developments in robotics even through investment in robotics by other sectors, such as the health sector where robot-assisted neurosurgery can reach very small areas deep inside the brain due to the accuracy of the latest robot technology, that are leading the way forward. Instead, automation solutions for apparel manufacturing will be dependent on new ways of thinking about how we can construct and join garments that can be assisted by robots using new developments in familiar technologies such as bonding, welding, or knitting whilst also re-imagining existing technologies. For example, 3D sewing technologies which were originally developed for the construction of composite materials in the 1980s are being revisited by key sewing machine makers such as PFAFF who presented at the 2022 Texprocess exhibition. In this example, the traditional sewing process is flipped, and instead of presenting the fabric to a sewing machine, the sewing machine is presented to the fabric assisted by a mobile sewing machine attached to the end of a robotic arm. This process has the potential to solve the issue of handling limp materials. But this potential solution like many others such as Sewbot's material-stiffening process needs considerable investment for further research and development, which can no longer be solely left to independents or the leading sewing machine makers such as PFAFF, JUKI, and Durkopp Adler, who have largely been the primary investors in automation R&D over the last 60 years. Now that key parameters to include new technology, new ways of thinking (McKinsey, 2018b), and the widespread willingness to achieve full automation appear to have aligned, perhaps the time is right for governments, the education sector, and those involved in apparel manufacturing to partner with machine makers to support and finance future research and developments in automation in the same successful way that DARPA and the Walmart Foundation have supported Software Solutions. This point is supported by Postlethwaite (2022), who argues that 'large-scale financial support from Government and the private sector should be made available to deliver a fashion manufacturing industry fit for the C21st' (Postlethwaite, 2022, p. 3). Ultimately, the consequences for the timely adoption or rejection of collective action are vast not just for the industry but for the planet.

REFERENCES

Brogan, C (2022) *Bendy Robotic Arm Twisted into Shape with the Help of Augmented Reality* [online], www.imperial.ac.uk/news/234362/bendy-robotic-twisted-into-shape-with/ (Accessed 03/01/2024)

FK Group (2022) *Augmented Reality Camera* [online], https://fkgroup.com/products/arc-augmented-reality-camera/ (Accessed 03/01/2024)

Fraser, K (2017) *The Impacts of a Sewing Robot, Fashion United* [Online], https://fashionunited.uk/news/business/the-impacts-of-a-sewing-robot/2017090125695 (Accessed 03/01/2024)

Ghobakhloo, M (2020) Industry 4.0, Digitization, and Opportunities for Sustainability. *Journal of Cleaner Production*, 252 [online], www.sciencedirect.com/science/article/pii/S0959652619347390 (Accessed 03/01/2024)

Gries, T and Lutz, V (2018) Application of Robotics in Garment Manufacturing. In: Nayak, R and Padhye, R (eds.) *Automation in Garment Manufacturing*, Woodhead, Elsevier, Cambridge

Hayes, SG and McLoughlin, J (2008) Technological Advances in Sewing Garments. In: Fairhurst, C (ed.) *Advances in Apparel Production*, Woodhead, Cambridge

IBM (2022) What Is Automation? *IBM* [online], www.ibm.com/topics/automation#:~:text=Related%20solutions-,Overview,where%20human%20input%20is%20minimized (Accessed 03/01/2024)

IFR (2022) *International Federation of Robots* [online], https://manplusmachines.com/differences-robots-cobots/ (Accessed 03/01/2024)

Intel (2022) *Different Types of Robots, How Robotics are Shaping Today's World* [online], www.intel.com/content/www/us/en/robotics/types-and-applications.html (Accessed 03/01/2024)

Jana, P (2018) Automation in Sewing Technology. In: Nayak, R and Padhye, R (eds.) *Automation in Garment Manufacturing*, Woodhead, Elsevier, Cambridge

Jones, RM (2006) *The Apparel Industry*, 2nd ed., Blackwell, Oxford

JustStyle (2020) *Micro-Factories—The Future of Fashion Manufacturing?—Just Style* [online], www.just-style.com/features/micro-factories-the-future-of-fashion-manufacturing/ (Accessed 03/01/2024)

Kalypso (2020) *The 2020 Digital Product Creation Survey Briefing* [online], https://kalypso.com/files/docs/Exec-Summary-Annual-Retail-Innovation-Adoption-Survey-2020_2020-12-17-215126.pdf (Accessed 27/12/2023)

Kuka (2022) *Cobots: The Intelligent Robot as a Colleague* [online], www.kuka.com/-/media/kuka-downloads/imported/87f2706ce77c4318877932fb36f6002d/kuka-cobots-en.pdf?rev=42bbb7a57ace485fab1de1ee834a2ec3&hash=C90426BB5E8F094678D728487F398EBE#:~:text=Working%20side%2Dby%2Dside%20with,rapidly%20changing%20production%20require%2D%20ments (Accessed 03/01/2024)

Machova, K (2018) Automation versus Modelling & Simulation. In: Nayak, R and Padhye, R (eds.) *Automation in Garment Manufacturing*, Woodhead, Elsevier, Cambridge

McKinsey (2018a) *Is Apparel Manufacturing Coming Home?: Nearshoring, Automation and Sustainability—Establishing a Demand-Focused Apparel Value Chain* [online], www.mckinsey.com/~/media/mckinsey/industries/retail/our%20insights/is%20apparel%20manufacturing%20coming%20home/is-apparel-manufacturing-coming-home_vf.pdf (Accessed 03/01/2024)

McKinsey (2018b) *Skill Shift: Automation and the Future of the Workforce* [online], www.mckinsey.com/featured-insights/future-of-work/skill-shift-automation-and-the-future-of-the-workforce (Accessed 03/01/2024)

McKinsey (2019) *The State of Fashion* [online], www.mckinsey.com/~/media/mckinsey/industries/retail/our%20insights/the%20state%20of%20fashion%202019%20a%20year%20of%20awakening/the-state-of-fashion-2019-final.ashx (Accessed 03/01/2024)

McLoughlin, J and Mitchell, A (2013) Mechanisms of Sewing Machines. In: Jones, I and Stylios, GK (eds.) *Joining Textiles, Principles and Applications*, Woodhead, Cambridge

Microdynamics (1981) *Microprocessor Control System for Sewing Machine, Google Patent* [online], https://patents.google.com/patent/US4359953A/en (Accessed 03/01/2024)

Moll, P (1997) Integrated 3D Sewing Technology and the Importance of the Physical and Mechanical Properties of Fabrics. *International Journal of Clothing Science and Technology*, 9(3): 249–251 [online], www.proquest.com/docview/228269343?parentSessionId=AtfmQY9LctrNl0nN6%2FHecsESfB6EoJ3ZlRgGezkVXKM%3D&sourcetype=Scholarly%20Journals (Accessed 03/01/2024)

Petro, G (2018) *See-Now-Buy-Now Is Re-wiring Retail* [online], www.forbes.com/sites/gregpetro/2018/01/31/how-see-now-buy-now-is-rewiring-retail/?sh=12a925442c0b (Accessed 03/01/2024)

Postlethwaite, S (2022) *Reshoring UK Manufacturing with Automation, Royal College of Art & Innovate UK KTN/Made Smarter* [online], https://iuk.ktn-uk.org/wp-content/uploads/2022/03/Reshoring-UK-Garment-Manufacturing-with-Automation-Thought-Leadership-Paper-final-2.pdf (Accessed 03/01/2024)

Revell, T (2017) *AI Will Be Able to Beat Us at Everything By 2060: New Scientist* [online], www.newscientist.com/article/2133188-ai-will-be-able-to-beat-us-at-everything-by-2060-say-experts/ (Accessed 11/12/2023)

The Robot Report (2019) *Adidas Closing Automated 'Speedfactories' in Germany, US* [online], www.therobotreport.com/adidas-closing-german-us-robot-speedfactories/ (Accessed 03/01/2024)

Rozelle, S, et al. (2020) *Moving Beyond Lewis: Employment and Wage Trends in China's High- and Low-Skilled Industries and the Emergence of an Era of Polarization* [online], https://link.springer.com/article/10.1057/s41294-020-00137-w (Accessed 03/01/2024)

Sewbo (2022) *Announcing the World's First Robotically Sewn Garment* [online], https://made-to-measure-suits.bgfashion.net/article/15295/47/Announcing-the-Worlds-First-Robotically-Sewn-Garment (Accessed 03/01/2024)

Song, C, et al. (2022) A Review on Three-Dimensional Stitched Composites and Their Research Perspectives. *Composites Part A: Applied Science and Manufacturing*, 153 [online], www.sciencedirect.com/science/article/pii/S1359835X21004449 (Accessed 03/01/2024)

Speer, J (2019) *Microfactories, Automation, Talent, Investment: Deconstructing the Reality of Made-in-the-USA Manufacturing* [online], https://risnews.com/microfactories-automation-talent-investment-deconstructing-reality-made-usa-manufacturing (Accessed 03/01/2024)

Tilley, J (2017) *Automation, Robotics, and the Factory of the Future* [online], www.mckinsey.com/capabilities/operations/our-insights/automation-robotics-and-the-factory-of-the-future (Accessed 03/01/2024)

Tyler, D (1989) *The Development Phase of the Textile/Clothing Technology Corporation Apparel Automation Project* [online], www.emerald.com/insight/content/doi/10.1108/eb002946/full/html (Accessed 03/01/2024)

Tyler, D (2009) *Carr and Lathams Technology of Clothing Manufacture*, Blackwell, London

Universal Robots (2022) *How Universal Robots Sold the First Cobot* [online], www.universal-robots.com/about-universal-robots/news-centre/the-history-behind-collaborative-robots-cobots/ (Accessed 03/01/2024)

Varshney, N (2021) *Connected Factories or Robotic Automation: What to Choose and Why?* [online], https://apparelresources.com/technology-news/manufacturing-tech/connected-factories-robotic-automation-choose/ (Accessed 03/01/2024)

6 Recent Advances in 3D Body Scanning and Measurement Systems

6.1 INTRODUCTION/SYNOPSIS

This chapter will open by providing an overview of the development of 3D body scanning. A detailed timeline will provide a summary of key developments associated with 3D body scanning, highlighting the concurrent trend of large national size surveys in the preceding two decades that were undertaken to redress the issue of poor fitting mass-produced apparel. The chapter will move on to consider the trend in the use of body measurement databases and data acceleration modelling for determining enhanced fit parameters and new product market. The chapter will also outline the types of 3D body scanners, including virtual app-based scanners and new measurement systems such as body volume index that exist in the market today and their relevance in supporting retailers to develop better fitting apparel whilst also meeting consumer demand for personalised and made-to-measure apparel. Avatar psychology and data privacy will also be addressed in relation to the use of scanners by the consumer. The chapter will close by considering advancements in 3D body scanning for ergonomic design and associated new technologies that will enhance 3D body scanning and digital measurements in the future. To provide the most current insight regarding recent and emerging advancements in 3D body scanning and digital measurement systems, key contributors include Richard Barnes (CEO and Founder of Select Research) who has pioneered Shape GB and the use of Body Volume Index, Richard Allen (Shape Analysis), Alex Chung, and Matthew McMillion (Artec 3D) along with leading experts and CEOs from three of the world's leading body scanning developers, including Dr Mike Fralix (TC2), Dr Helga Gaebel (Avalution: Human Solutions/Humanetics), and David Bruner (Size Stream).

6.2 3D BODY SCANNING

Accurate body measurement data is a key requirement for good design and fit. Manual measuring methods that use traditional tape measures are generally time-consuming, invasive, and can compress the surface of the skin often producing inaccurate and variable results. Alternatively, 3D body scanners automate the measurement process by capturing highly accurate body dimensions in just a few seconds and do not require any contact. Repeatability of measurement has been proven to be more accurate compared to manual measurement (Istook, 2008). The 3D body image can be captured using a variety of technologies such as laser

DOI: 10.1201/9781003126454-6

technology, depth sensors, structured light, millimetre/microwaves, infrared waves, or photogrammetry (Zeraatkar and Khalili, 2020). Each of these methods has their own advantages, limitations, and associated costs depending on the design of the scanner, the sophistication of software, and the proposed end use of the scanned data.

In addition to image capture technology, dedicated software is also needed to enable scanned data to be accessible for analysis. Data is output in the form of a *point cloud* which is transformed using sophisticated algorithms facilitating 3D avatar generation to enable analysis and a measurement extraction profile, meaning virtually any part of the body can be measured when required (Elbrecht and Palm, 2014). This detailed measurement data can be saved in a number of formats and exported to a variety of CAD systems for the purposes of product development and digital 3D prototyping whilst also supporting the development of personalised and bespoke apparel. Measurement data can also be used for a variety of analytical and visualisation purposes for health, fitness, as well as marketing and e-commerce-related activities.

6.2.1 Impact of Human Physiology on 3D Body Scanning

A key requirement for achieving highly accurate 3D body scan data relates to the need to eliminate even small movements such as breathing and blinking during the scan process, as this can compromise the accuracy of measurements. According to Kaneko (2014), 'the breathing amplitude in healthy humans is on average one-third of a deep breath approximately 3 cm, and this can be variable from person to person' Kaneko (2014). Ultimately, these displacements can result in errors in the resulting 3D model. Some scanning companies recommend capturing three scans and then taking the average reading to compensate for these natural movements. With this in mind, recent advances by the leading 3D body scanning companies have resulted in the development of highly accurate body scanners that can scan a body in just one second with 1 mm of accuracy.

6.2.2 Development of 3D Body Scanning Technology

Body scanning technology is still relatively new, and many early developments were largely centred within the UK and the United States. Trailblazers included collaborations between UK-based apparel retailer M&S and Loughborough University in the late 1980s, as well as Cyberware, who were key developers of 3D scanners for the film and gaming industry in the United States. Cyberware scanning technology was also adopted at the time by the US military for improving the fit of military uniform. Other early adopters for the purpose of mass customisation in the late 1990s included Levis Strauss and later in 2001 Brooks Brothers in the United States. This led on to the establishment of National Size Surveys around the world such as Size UK in 2000, which used the highly accurate 3D body scanning booths developed by the American scanning company TC². The following timeline provides a summary of key developments associated with 3D body scanning and highlights the concurrent trend of large national sizes surveys that have been carried out during the last 20 years for the purpose of redressing the issue of poor fitting mass-produced apparel.

TABLE 6.1
Timeline of 3D Body Scanning Technology

1964	Vietorisz used light source and photodetectors to measure the human silhouette.
1977	Clerget, Germain, and Kryze measured objects with a scanning laser beam.
1985	David and Lloyd Addleman developed scanning laser beam (became Cyberware).
1987	Cyberware developed scanners for scanning the head, face, and feet for computer graphics.
1987	Loughborough anthropometric scanner was commissioned by M&S.
1994	Levi Strauss introduced 3D body scanners to store and commence mass customisation programmes; the first was called 'Personal Pair' and second was called 'Original Spin'.
1995	Cyberware whole body scanner (also used by US Air Force to improve uniform fit).
1996	UK defence clothing and textile agency (DCTA) and National Engineering Lab Scotland developed 3D measurement system known as Auto-Mate.
1998	TC2 launched the first 3D body scanner for the apparel industry, used by Levi Strauss, US Navy, North Carolina State University, and Clarity Fit tech.
2000	Size UK—using TC2 3D body scanners.
2001	Brooks Brothers introduced TC2 3D body scanner to store-to-support customisation
2002	David Lloyd Leisure Centres introduced 3D body scanners to track changes to body shape.
2002	Size USA using TC2 3D body scanners.
2004	Size Mexico using TC2 3D body scanners.
2007	Size Spain—Avalution: Human Solutions.
2008	Size Germany—Avalution: Human Solutions.
2012	Size Italy—Avalution: Human Solutions.
2012	Size Stream company established (3D body scan business established by the Tal group, a large global manufacturer specialising in innovation in product development)

Most notable is the 2011 National Childrenswear Survey in the UK which demonstrated that UK children's body shape had dramatically changed since the last published data from the British Standards Institute (BSI) in 1990. At the time, this evidence assisted to reinforce to key industry retailers the need for a new UK industry standard for fit. The four sponsoring retailers, including Next, Monsoon, Shop Direct, and George at Asda, were able to use the data to collaboratively harmonise their measurement standards and vastly improve the fit of UK children's clothing. Also, they supported the British and European clothing industry by making the original 3D body scan measurement data available to other retailers and organisations (Aston, 2011).

More recently, go-to style national size survey has become less popular for a number of reasons. First, they are extremely expensive to organise and deliver, and most rely on corporate or government sponsorship to support participant engagement. Second, critics argue that sizing surveys are not effective because most mass-produced clothing still does not fit, as evidenced by high returns rates via e-commerce-based shopping (Cathcart, 2020). Third, new and emerging developments in virtual scanning technology now enable consumers to scan and submit data to their preferred brand themselves from the privacy of their home. Fourth, many fashion businesses are realising that it may be more effective to focus on how to fit or hone into a much narrower consumer base including the individual rather than how to fit the masses, and this approach has been successfully realised through the creation of body measurement databases.

6.2.3   BODY MEASUREMENT DATABASES AND DATA ACCELERATION MODELLING (DAM)

Body measurement databases enable the statistical analysis of data concerning the distribution of body shape and fit. Other parameters such as changes in diet and lifestyle, health, fitness, and ethnicity can be used to both inform and enhance apparel fit and in the past have required regular data gathering, mostly via large national size surveys to keep them up to date. An example here includes one of the global leaders in body measurement databases known as Avalution, part of the Human Solutions group who specialise in 3D body scanning. Since 2012, they have undertaken size surveys in most major countries in the world to include the United States, Brazil, France, Sweden, Poland, Romania, China, India, and South Africa. This has led to the development of the world's most comprehensive body measurement database called the iSize Portal©. The portal enables fashion brands to analyse statistical data which they can also use to generate avatars to support body shape analysis, and this can be used to target new consumers and new markets. Data can be mined from parameters such as size, shape, age, gender, as well as social demographic data, including buying behaviour and sports activity. Data searches of new countries can also assist in evidencing, for example, the need to increase leg length for an existing style, and projections can be made on how this change might increase market share in this potential market simply by fulfilling the needs of the target market. This iPortal data is also attached to a mobile scanning app, which they call The Virtual Scanner. This has been developed for the requirements of the fashion retail industry as it enables the use of customer-specific measurements to recommend the best-fitting size for each individual. This is especially useful where multiple fit variations such as high, medium, or low rise etc. are concerned. The advantages of this technology for online purchases are that it also includes virtual try-on with 3D-simulated products in web shops.

One of the major limitations with body measurement databases is that their lifespan can often be relatively short as national demographics can quickly change even within a five-year span. Many databases are now able to compensate for change and are equipped with Data Acceleration Modelling (DAM). In the case of the iPortal database, the importance of DAM was originally realised having been previously developed for the automotive industry where the product development timeline is much longer, involving the development of car interiors including car seats which require specific anthropometric data that enables the prediction of how people might change in the future. The ability to undertake acceleration modelling therefore solves the issue of size surveys going out of date so quickly and is an important consideration for any new large-scale body-scanning project.

Projects similar to iPortal have also been developed in the UK, and recent data from the UK Office for National Statistics makes the case for this work more compelling as world migration to the UK continues to accelerate, meaning its demographics have greatly changed in less than five years. According to the ONS (2020), migration from the EU has fallen significantly since 2016, while migration from elsewhere has increased. This change has been recognised by the key apparel retailers in the UK and has led to the development of a new approach to anthropometry such as Body Volume Index as discussed later in this chapter.

6.3   NEED FOR DIFFERENT 3D BODY SCANNERS

3D body scanners are not the same and each has their own advantages, limitations, and associated costs depending on the design of the scanner, the sophistication of software, and the proposed end use of the scanned data. Often, lower spec scanners have found new markets. For example, since 2002, David Lloyd leisure centres introduced 3D body scanners to track changes to body shape. Sector interest in this technology has since continued to boom. Some companies such as Styku, who originally focused on developing 3D body scanning equipment for the apparel industry, have also opted to focus their efforts on developing lower cost scanners for the lucrative health and fitness sectors where the accuracy and speed of the technology are not so crucial for recording changes in body shape associated with health, fitness, and sport. Often, these scanners operate at much slower times, where scan times can take up to 35 seconds compared to the cubicle scanners that can scan in just one second. This is because these scanners are often based on a turntable design which rotates through a fixed scan point, so speed needs to be curtailed for safety purposes. Many new designs have handles incorporated to help steady the participant, minimise movement, and improve scan quality.

Alternatively, if developing technical performance apparel or undertaking national size surveys, a booth-style laser scanner will provide the highest degree of speed and accuracy such as the Sizestream 2020 model or the Avalution Vitus Scanner. Both examples are currently considered as gold standard 3D body scanners delivering (+/-1 mm) scanning accuracy.

In contrast to the latter, retailers who offer consumers a predictive sizing service might utilise an intermediate scanner such as the Avalution AVAone scanner utilising depth sensor technology. In this particular scenario, the degree of accuracy will be approximately +/-5 mm but accurate enough for this end use. This category of scanner often known as the 'fitting room scanner' because of their in-store location, is being increasingly sourced by retailers pursuing made-to-measure (MTM) apparel. This service has an additional benefit for the retailer, as it enables measurement databases to be developed to inform future sizing, which will be discussed further later in the chapter.

6.4   BODY SCANNING FOR BESPOKE AND
MADE-TO-MEASURE/MASS CUSTOMISATION

3D body scanning booths are currently still regarded as the go-to technology when developing both bespoke apparel such as individually tailored suits where each design

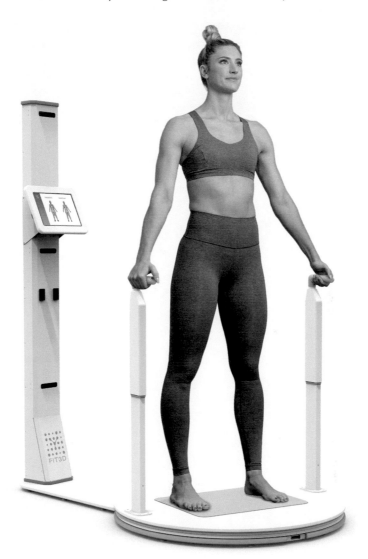

FIGURE 6.1   Fit3D scanner.

Source: Courtesy of www.fit3d.com

and fit is unique; as well as custom-fitting apparel also known as made-to-measure (MTM). This approach was originally pioneered by companies such as Levi Strauss in the 1990s when it integrated a 3D body scanner into its Union Square Store in San Francisco to deliver customised Levi's jeans through its 'Original Spin' programme. In 2001, Brooks Brothers introduced a TC scanner for its bespoke tailoring, followed by Selfridges in London, and Le Bon Marché department store in Paris. This trend filtered down to the high street when fashion brands such as New Look launched a 3D body scanning programme with Bodymetrics throughout the early millennium period.

FIGURE 6.2   Avalution AVAone scanner.

Source: Courtesy of www.avalution.com

Appetite for MTM apparel began to curtail from 2004 onwards as concurrent trends in offshore sourcing decimated localised manufacturing in the West. Levi's, for example, closed its last UK-based manufacturers in 2004. However, in the last ten years, MTM apparel has gained momentum in response to sustainability agendas and customer demand. Levi's, for example, revisited mass customisation in 2014 through the launch of an exclusive MTM denim jean programme called *Lot No. 1*. Each pair of jeans is individually made by highly experienced in-store tailors. At the other extreme, the high-street fashion brand H&M has recently invested in MTM in a similar way having installed body scanners in its Weekday stores to facilitate custom-fit made-to-order jeans.

Other examples include London-based start-up Formcut, owned by Tal, one of the worlds' largest apparel manufacturers. It specialises in MTM menswear enabled by state-of-the-art 3D body scanners and innovative app-based virtual scanners developed by one of the leading developers of scanning technology called Sizestream, which it also own.

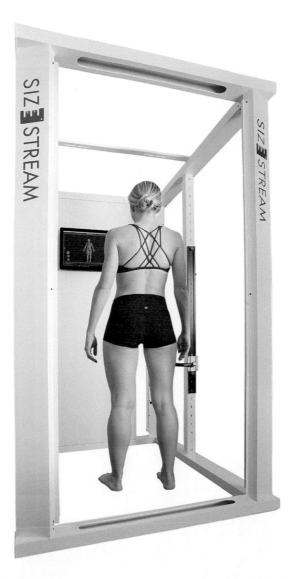

FIGURE 6.3   Sizestream 3D body scanner—1-mm accuracy/8-second scan time.

Source: Courtesy of Copyright Size Stream LLC, Cary, NC USA

6.5 DIGITAL MEASUREMENT SYSTEMS: VIRTUAL SCANNERS AND BODY VOLUME INDEX

Growth of e-commerce and demographic shift have heightened the need for brands to find new technologies to improve size and fit. For the e-commerce sector, poor fitting clothing has meant many consumers order multiple sizes to find the best fit. This in turn results in unavailable stock, large overheads associated with processing customer returns, which are currently estimated to be around 60% (Charlton, 2020). Ultimately, fewer garments are sold at full price, with increasingly more assigned to

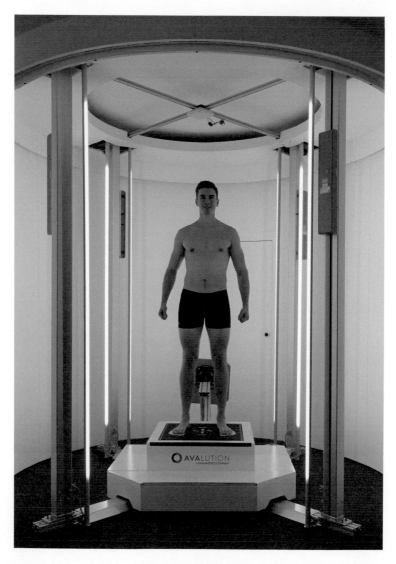

FIGURE 6.4   Vitus3D body scanner—1-mm accuracy/6-second scan time.

Source: Courtesy of www.avalution.com

sale racks and most ending up in landfill sites. Empowering the customer to capture and share their own personal measurement data, using new and emerging Virtual Scanning Technology (VST) can assist to redress these issues.

Virtual scanners such as those developed by Avaluation, part of the Humanetics group, are designed as a specialist app on a mobile phone and require users to input basic data such as gender, size, and age to create a representative avatar for an individual person or a target group (Avalution, 2021). This type of technology has also been used to deliver a UK national size survey called Shape GB (Roberts, 2018). Unlike traditional national surveys, participants do not need to physically attend a road show event to have their measurements taken. Instead, they are able to capture their own fit and shape data at home using the Shape GB app on their mobile phone. The app utilises a new revolutionary approach for calculating the size and shape of the body in 3D called Body Volume Index (BVI). Originally launched in 2017, BVI is also being adopted by many of the world's health authorities in favour of Body Mass Index (BMI). Body measurements are obtained by combining a person's 3D shape via a front and side photo captured and submitted by the consumer, with data on height, weight, age, gender, and ethnicity. The Shape GB app adds body measurements to a secure national sizing database, which utilises powerful algorithm-based software developed by the company TC², the same company responsible for some of the world's original national size surveys. This software is capable of calculating the mean value for standard size clothing for a target customer group. Ultimately, the Shape GB survey not only captures consumers' measurements, it can also map their shape, providing UK retailers with comprehensive data to develop better fitting apparel. At the time of writing, the survey has not been completed due to the COVID-19 pandemic, and therefore the success and impact of this project cannot yet be fully reported.

Developments in virtual scanning technology have further accelerated since 2020 well ahead of predictions made in 2019 by Levi Strauss' CEO Chip Bergh, who stated that he believed clothing sizes will disappear within ten years, and customers will scan themselves and buy clothing that fits their individual form. This has already become a reality, and in less than 12 months, brands such as Formcut have redeveloped their virtual scanning app which originally required a two-man capture, a specialist form-fitting suit, and a floor-based template. The new app is reported to be so good that the developers Sizestream have announced in April 2021 that they intend to discontinue the 'gold standard' booth scanner and move forward with just two app-based virtual scanners, one of which is aimed at the health and fitness sector called MeThreeSixty, and the other is aimed at the apparel industry.

Amazon, which acquired the tech company Body Labs in 2017, has also realised the potential of virtual scanners and in December 2020 launched an MTM service using a virtual scanning app called 'Made for You'. Currently, the range includes a basic T-shirt, and the app requires customers to supply two photos plus their height and weight. It is mooted that Amazon is currently planning further developments in the area of 3D body scanning and modelling to deliver better fitting apparel in the future.

Ultimately, the gold standard 3D body scanner remains an indispensable tool for measurement capture for accurate avatar creation. Likewise, interesting combinations of the booth scanner and the virtual scanner are emerging on the high street. H&M, for example, is in the process of piloting a 'virtual fitting solution' from June 2021.

This scheme enables customers to be scanned in-store, and in return, they receive a digital twin of themselves that can be used in conjunction with a dedicated app for virtual fitting of any style at home (Husband, 2021). This approach not only increases sales and streamlines online sales; it also allows brands to develop a comprehensive

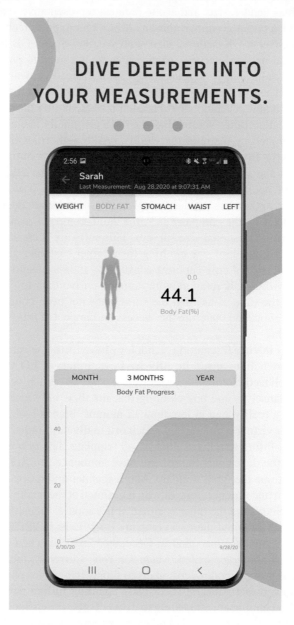

FIGURE 6.5   MeThreeSixty Sizestream Scanner App.

Source: Courtesy of Copyright Size Stream LLC, Cary, NC, USA

anthropometric database to inform the brands' overall fit strategy. H&M plans to launch its digital fitting rooms in selected stores in Germany from June 2021.

6.6   AVATAR PSYCHOLOGY AND DATA PRIVACY

It is a well-known fact that most people view their personal avatars negatively, and there have been emerging ethical and psychology-related concerns regarding the release of the avatar to the customer especially in the race to enable virtual try-ons using self-captured avatars. Body scanning provides an accurate silhouette of an individual's shape; however, most people are negatively surprised when they first view their avatar. Many comment how it does not look like them, and that they look shorter and wider in most cases. According to Mindless (2020), whilst better fitting apparel may assist with positive body image attitudes, some studies suggest a 3D scan of your body may negatively affect mood and lower self-esteem leading to scrutiny of body image shape. With this in mind, some companies manipulate the avatar image but the measurements remain the same whilst others aim to help users set health targets. For example, Sizestream has introduced a slider feature within their MeThreeSixty virtual scanner that enables users to view a current and future avatar side by side, simulating a new personal target whilst an in-built safety feature prevents individuals setting extreme weight-changing goals.

Many experts involved in body scanning agree that body scanning is facing increased legal regulations regarding data protection. Current data privacy laws mean that the end user owns the data and can decide to share their data in exchange for customised goods and services from other companies such as fashion brands. So, for example, if an end user wants to try on clothing with a particular brand, they need to give permission to share their data which is usually a photo plus measurements and sometimes, additional demographic information. Sharing of data is enabled using a QR code or hyperlink, which will enable the company to upload their data. Each time data is uploaded by a third party, the company which owns the virtual scanning technology will receive a small payment. The current rate is a minimum of US$1 per upload, but this can easily mount up if consumers decide to share across multiple channels. Further clarity and consumer reassurance of where such data may ultimately end up are needed especially as consumers are currently providing photos of themselves in form-fitting apparel. A related new feature undergoing current development is 'no-photo upload' whereby the photo can be analysed on the end user's phone but is never uploaded by the app or third-party company. This means in future children's sizing may be supported with this technology. This development also negates the issue of some users wearing form-fitting apparel that is too loose meaning accurate measurements cannot be captured.

6.7   ADVANCEMENTS IN 3D BODY SCANNING FOR ERGONOMIC DESIGN

A long-standing limitation of 3D body scanning technology is that some areas on the body do not scan well. These areas are also often covered with underwear and can result in a deformation of the body particularly at the crotch, breast point, and gluteal fold, and they are therefore usually adapted post scanning to assure accuracy of fit. Until recently, the vast majority of 3D body scanners required participants to stand in the upright position, meaning positional scans such as in a cycling or seated pose were not possible. However, recent developments such as the 'Vitus' Avalution/ Human Solutions scanner (part of the Humanetics Group) can generate scans in the seated position assisted by powerful algorithm software. This capability was originally developed for the automotive industries to create virtual vehicle interiors and custom 3D CAD avatars that realistically simulate vehicle occupants to evaluate ergonomics and comfort parameters during the early stage of prototype development but has now also been utilised within the apparel industry for developing performance or industrial apparel designs based on seated stances.

In relation to body scanning for apparel product development and innovation involving different body poses, one of the most impressive advancements is the Artec Leo Scanner. This revolutionary scanner has very recently facilitated the development of performance apparel for cycle Olympians and has culminated in gold, silver, and bronze medals in the 2021 Olympic Games. Further details on this process are provided in a case study later in the book. The Artec Leo Scanner has inbuilt software utilising AI technology that enables a huge variety of objects to be scanned within a wide range of distances and is therefore applicable across a varied range of sectors. Likewise, its wireless hand-held design enables enhanced flexibility and accuracy during the scanning process. For example, it has recently been used by

FIGURE 6.6 Automotive product development to evaluate ergonomics and comfort parameters during the early stage of prototype development.

Source: Courtesy of www.avalution.com

NASA for the development of spacesuits for the Artemis program and also by the film, gaming, and Formula One racing industries to create award-winning outcomes.

The scanner uses VCSEL (Class 1 Laser) infra-red structured light and has two front lenses. One of the lenses emits structured light onto the target, and the other is made up of 12 LEDs and two cameras that work to accurately capture image, colour, and texture. The scanner is equipped with sophisticated sensors to facilitate self-positioning in space.

The Artec Leo scanner is also one of the first cordless hand-held 3D scanners with inbuilt automatic processing. Objects are scanned using a spray-painting style motion, which is captured, processed, and viewed in a touchscreen window in real time without the need for additional cables and computers. Whilst scanning you can rotate the model in the touchscreen to check all areas have been captured and can easily rescan areas to fill in missing gaps. Areas such as the crotch and underarm can be accurately scanned, eliminating the need for additional software traditionally used to fill or *clean* missing gaps or inaccuracies post scanning.

6.8   THE FUTURE

In terms of the future of 3D body scanning and digital measurement systems, one of the most recent and impressive technologies that assists in transforming the quality of virtual scanners is LIDAR, which has already been included on many high-spec mobile phones.

LIDAR stands for Light Detection and Ranging and works by emitting a laser to the environment and measuring the time it needs to bounce back into the sensor to calculate the distance of each point. LIDAR technology has existed since the 1960s,

FIGURE 6.7   Artec Leo scanner.

Source: Courtesy of www.artec3d.com

first used in military aircraft, and better known with its association with Apollo 15 and its use for mapping the surface of the moon. In recent years, it has been incorporated into driverless vehicles to assist in the detection of pedestrians and has been used in robot vacuums and lawnmowers to assist such devices with self-positioning. This technology will greatly assist to revolutionise virtual try-on and virtual scanning technology for the apparel industry and ultimately reduce returns and waste associated with the age-old issue of poor fitting apparel.

REFERENCES

Avalution (2021) *Virtual Scanners* [online], www.avalution.net/en/fashion/3d-body-scanner/index.html (Accessed 02/01/2024)

Cathcart (2020) The Body Shaped Hole in Your Digital Transformation. *The Interline* [online], www.theinterline.com/2020/04/09/the-body-shaped-hole-in-your-digital-transformation/ (Accessed 02/01/2023)

Charlton, G (2020) *Ecommerce Returns: 2020 Stats and Trends* [online], www.salecycle.com/blog/featured/ecommerce-returns-2018-stats-trends/ (Accessed 02/01/2024)

Elbrecht, P and Palm, KJ (2014) *Precision of 3D Body Scanners.* INES 2014, IEEE 18th International Conference on Intelligent Engineering Systems, 3rd to 5th July 2014, Tihany, Hungary [online], https://ieeexplore.ieee.org/stamp/stamp.jsp?tp=&arnumber=6909355&tag=1 (Accessed 06/01/2021)

Husband, L (2021) *H&M Launches Virtual Fitting Room Experience in Germany* [online], www.just-style.com/news/hm-virtual-fitting-room-germany/?cf-view&cf-closed (Accessed 02/01/02024)

Istook, CL (2008) 3D Body Scanning to Improve Fit. In: Fairhurst, C (ed.) *Advances in Apparel Production*, Woodhead and CRS Press, Cambridge

Kaneko, H (2014) Estimating Breathing Movements of the Chest and Abdominal Wall Using a Simple, Newly Developed Breathing Movement-Measuring Device. *Respiratory Care*, 59(7): 1133–1139. https://doi.org/10.4187/respcare.02778 (Accessed 02/01/2024)

Mindless.mag (2020) *Fashion and Psychology: The Future of Fashion: How 3D Body Scanning Could Impact Consumers* [online], www.mindlessmag.com/fashion-and-psychology (Accessed 02/01/2024)

ONS (2020) *How the UK's Population Has Changed Since the Start of the 20th Century, Our Population, Where Are We? How Did We Get Here? Where Are We Going?* [online], www.ons.gov.uk/peoplepopulationandcommunity/populationandmigration/populationestimates/articles/ourpopulationwherearewehowdidwegetherewhereare wegoing/2020-03-27 (Accessed 27/12/2023)

Roberts, L (2018) Leading Fashion Retailers Back Major UK Wide Sizing Survey. *The Industry Fashion* [online], www.theindustry.fashion/leading-fashion-retailers-back-major-uk-wide-sizing-survey/ (Accessed 27/12/2023)

Zeraatkar, M and Khalili, K (2020) *A Fast and Low-Cost Human Body 3D Scanner Using 100 Cameras* [online], www.researchgate.net/publication/340566806_A_Fast_and_Low-Cost_Human_Body_3D_Scanner_Using_100_Cameras (Accessed 02/01/2024)

7 Recent Advances in Digital Pattern Making Systems

7.1 INTRODUCTION

This chapter will provide an overview of the development of digital pattern cutting systems, outlining how major developments in both processing speed and graphic processing power have ultimately enabled a merging of 2D and 3D digital pattern systems. The chapter will move on review recent and emerging technologies in this area, focusing on one of the world's leading providers of digital pattern cutting systems Lectra and how they are meeting current industry challenges that include the drive towards sustainable design practices and sustainable supply chains, faster fashion, creating clothing that fits, and offering consumers options for personalised apparel by developing the potential for fully automating pattern cutting practices and processes during the product development and production development stages of apparel generation.

To provide the most current insight regarding recent and emerging advancements in this area, the key contributor includes Mathieu Bonnenfant, Vice-President of Product Marketing, Lectra.

7.2 DEVELOPMENT OF DIGITAL PATTERN CUTTING SYSTEMS

Prior to the invention of the computer and up until the 1980s, patterns used by the apparel industry were cut by hand using basic equipment such as paper, card, pencils, and a variety of tools such as rulers, pattern drills, notchers, and tape measures. Markers for bulk production were manually created by tracing around graded patterns that had been transferred to a thick card.

In response to complications associated with the globalisation of apparel production in the latter part of the twentieth century, digital pattern making systems were often developed alongside computer-aided design systems (CAD) some of which incorporated Product Lifecycle Management (PLM) capabilities. As well as enabling speedy and accurate pattern manipulation, they also improved efficiencies by streamlining the design, development, and manufacture of apparel by enhancing collaboration and communication across the apparel supply chain. Initial developments of digital pattern cutting systems commenced in 1980, when Gerber Technology acquired the AM-1 computer-aided design system from Hughes Aircraft Company (Baytarm and Sanders, 2020). This was the basis for a next-generation apparel CAD system, the Gerber AM-5. In 1976, Lectra introduced its first digital pattern grading and marker making solution called LS Model and LS Mark. In

DOI: 10.1201/9781003126454-7

1996, Lectra completely revamped its CAD solutions, with the launch of Modaris and Diamino Fashion along with the freeline pattern drafting table. Modaris is still a leading pattern making solution today. Meanwhile, Gerber Technologies continued to develop a pattern making, grading, and marker system and launched AccuMark® in 1988 and by 1994 revolutionised pattern design with the introduction of the digital Silhouette™ pattern drafting table. During this period, both the hardware and the software associated with digital pattern cutting were expensive as were the costs associated with training. For these reasons, digital pattern cutting systems were generally adopted only by wealthy apparel manufacturing businesses such as Dewhirst in the UK, who supplied high-street retail giants of the time such as M&S. This situation has recently changed, as key developers of digital pattern cutting systems such as Gerber Technology and Lectra have made their software accessible to all using software as a service (SaaS) model.

From the 1990s, the digital pattern cutting arena for apparel was dominated by three giants including American-based Gerber Technologies, Lectra which originated in France, and their Spanish rival Investronica Sistemas, which was acquired by Lectra in 2004. This strategic acquisition meant that Lectra was able to further develop and expand its apparel markets whilst maintaining its work with the automotive and furniture industries (Barrie, 2004a, 2004b).

Key developments during this period included the release of Gerber AccuNest™ in 2006. This innovation was remarkable because for the first time, markers could be generated automatically, and users needed limited training. Since the millennium, major developments in both processing speed and graphic processing power have ultimately enabled a merging of 2D and 3D digital pattern systems. Most notable was the release of Lectra's Modaris 3D software in 2007 and Gerber's AccuMark® 3D in 2015. Developments in the area of 3D have continued, and in 2018, Gerber acquired Avametrics, enabling it to focus on cloth simulation technology to support high-quality 3D simulations of its products on customisable avatars.

Today, the digital pattern cutting arena is dominated by Lectra, which acquired its only rival Gerber Technologies in February 2021 for EUR300 million (Wright, 2021), concurrently releasing impressive new developments in digital pattern cutting technology software to meet the ever-changing challenges of the increasingly digitalised fashion industry. The following section will contextualise the current industry challenges driving demand for concurrent advances in digital pattern cutting systems.

7.3 CURRENT INDUSTRY CHALLENGES DRIVING DEMAND FOR DIGITAL PATTERN CUTTING SYSTEMS

The fashion industry is currently facing several key challenges, which include the drive towards sustainable design practice and sustainable supply chains, faster fashion, creating clothes that fit, and offering consumers options for personalised apparel. As a consequence of the recent pandemic, the accelerated movement towards a digitalised fashion industry has further highlighted the need for enhanced integration of design, development, and manufacturing processes as well as the need for seamless communication and increased automation. At one extreme, the fashion industry is currently working to improve sustainable practice throughout the supply chain, and at the other extreme, fast fashion continues to flourish and has become faster than ever in response to consumer demands. Trends are constantly changing at a rapid pace, as consumers increasingly adopt a see-now-buy-now mentality (JustStyle, 2018). Therefore, brands need to produce a larger number of small-volume collections with a greater variety of style options that are competitively priced and delivered within a shorter time frame to remain competitive. At the same time, changes in diet, lifestyle, and demographics have meant that many brands are producing clothes using fit parameters that have become out of date, and this has meant high return rates and elevated costs associated with the administration of returns. In response to this, and to meet the growing trend towards personalised apparel, some brands have diversified their business models to offer made-to-measure apparel that is produced using on-demand manufacturing methods (Lectra, 2021). Fully integrated digital pattern cutting systems are therefore a fundamental requirement in supporting these current industry challenges, and the most recent advances by Gerber Technology and Lectra, who are the main leaders in this field, will form the focus of the rest of this chapter.

7.4 RECENT ADVANCES IN DIGITAL PATTERN CUTTING—GERBER TECHNOLOGY

Gerber's highly sophisticated AccuMark platform is one of the only fully integrated digital pattern cutting systems in the world that enable the integration of 2D and 3D CAD with made-to-measure, marker making, and nesting software. The subscription-based software is accessible to companies of all sizes from top global brands to new start-ups, and its flexibility means it can assist in the rapid development of made-to-measure garments as well as managing the complicated workflows associated with mass production. Since its introduction more than 30 years ago in 1998, Gerber has continued to develop and perfect AccuMark in response to changing industry challenges, which include creating sustainable supply chains, faster fashion, creating clothes that fit, and offering consumers options for personalisation of apparel. The AccuMark platform includes AccuMark digital pattern cutting software, AccuMark 3D, AccuScan, and AccuPlan Made to Measure, all of which support the design and development of apparel, whilst AccuPlan and AccuNest support production planning and manufacture. The key features and latest advances for each of these digital tools will be discussed as follows.

7.4.1 ACCUMARK

The latest version of AccuMark offers several new features that assist to streamline pattern design. For example, it is now possible to consolidate fabric colourway data into one model, update artwork changes across multiple models, and identify marker information for digital printing. The *replace Image* function within AccuMark enables the processing of multiple image substitutions within one model, permitting the user to identify specific trim images on a piece and swap the placement within a marker for another image. The location, grade, and scale are maintained for each

FIGURE 7.1   The Replace Image function within Accumark.

Source: Courtesy of www.gerber.com

replaced image. One example is team sports, where the fabric and majority of trim images are consistent, but the player's name needs to be assigned for each team member. With Replace Image one model can be developed and maintained, and data about substitution name and sizes can be imported from the company's ERP system and utilised in Easy Order. Coupled with batch processing that now also support print file generation, the production process can be quickly automated.

The smart pattern making tool within AccuMark claims to reduce development time by two weeks by reducing errors. The grading process can also be automated with proportional and multidimensional grading, and it is also possible to compare grade, annotations, or piece shapes using a new copy-and-paste function, which enables patterns to be easily moved and compared from different locations. Detailed design data can be shared outside AccuMark using integration tools within Gerber's PLM system, YuniquePLM®.

7.4.2 AccuMark 3D

There are currently a variety of pattern-dependent 3D digital tools on the market that claim the ability to enable users to create more styles in less time, validate fit, facilitate seamless collaboration, and respond to sustainable design practice by reducing sampling and therefore reducing waste and carbon footprint. AccuMark 3D does all of this but has a leading advantage because it is supported by a long-established and globally tested fully integrated digital pattern cutting solution that enables the integration of 2D and 3D CAD with highly sophisticated made-to-measure, marker making, and nesting software, all of which are linked by a highly efficient PLM system.

7.4.3 AccuScan

AccuScan is a high-speed automated digitising system that enables physical patterns to be quickly and accurately digitised using a digital camera and a specialist

AccuScan mat imprinted with specialised targets. The targets enable the AccuScan software to interpret the scale and perspective of the pattern images and accurately convert the images to digital piece data for use in the AccuMark system. The mat can be hung on a wall or laid on a table, thereby consuming less floor space than traditional digitising equipment. Other benefits over manual digitising include the fact that it is easier to use and more ergonomic, it eliminates human error during the manual digitisation process, and is typically 20–50% faster than digitising manually. AccuScan can also automatically detect notches, grain lines, internal lines, drill holes, and the part perimeter, and surprisingly, patterns made from any type of material can be used to create AccuMark pattern pieces.

7.4.4 ACCUMARK MADE-TO-MEASURE

AccuMark Made-to-Measure (MTM) software is unique as it is currently one of the only software solutions providing true automation from pattern modification through advanced rule-based specifications and order creation. Some of the current features facilitate the alteration of a model based on an individual's style choices and measurements. You can also use measurements derived from high-tech 3D body scanners or low-tech tape measures and create AccuMark orders, models, and size codes based on unique customer requirements. For many companies offering personalised apparel, database management is becoming increasingly important, and the AccuMark MTM software enables the creation of a customisable database with smart search tools for querying orders by customer name or other preferred details. The software also reduces time to market by automating order selection and decision-making using powerful knowledge-based rules. It also automates *blue pencil* alterations as well as the batch processing and work-in-progress functions. Other automatic features include the potential to further streamline productivity and maximise material utilisation through the ability to generate and export cut data via its integration with AccuNest™ which means that markers can also be generated automatically.

7.4.5 ACCUPLAN

AccuPlan is a powerful spread and cut digital planning tool that can automate the lay-planning and marker making process. This fully automated solution incorporates all of the stages relating to cut planning through to nesting and ultimately reduces material/fabric consumption and labour costs and contributes to a reduction in production time. Sophisticated spread and cut plans can be produced that are based on a variety of factors such as cutting table dimensions/availability or other data relating to individual fabric lengths, including highly complicated information where thousands of fabric rolls for a single order are concerned. Material utilisations for whole orders can be run and checked in advance and therefore support product development and garment engineering to achieve highly efficient designs. Standardised nesting constraints can also be defined to support high-quality laying and cutting results. A production tracker feature enables the planning of multiple cut work orders as well as the tracking of orders through the cutting room by using digital tickets and barcodes.

7.4.6 AccuNest

AccuNest leverages automatic processing with powerful nesting algorithms to generate a variety of nesting possibilities using an iterative process, which ultimately results in the automatic selection of the model with the highest fabric utilisation. This assists in speeding up the marker marking process and therefore reduces labour cost. Specific constraints can be applied to pieces such as directional rotation, tilt allowances, and also spread-constraints such as avoiding nesting on material defects to ensure quality and minimise costly mistakes in the cutting room. Similarly, AccuNest can also automatically create markers for stripe and plaid materials whilst enabling specified matching requirements. The software can also group pieces by size or bundle into sections to maintain proximity of pieces in the marker for shade variations. Operators have the flexibility to control the placement of critical parts whilst also allowing the system to automatically nest remaining parts. Highly complex markers can be automatically nested as pre-marking rules can be input at the start. A recent feature called UltraQueue allows users to submit markers from AccuMark® to a holding queue for processing. Colour-coded management view of all work in process and completed jobs allows users to see at a glance which markers are in what stages without having to open every marker or run a job report.

7.5   RECENT ADVANCES IN DIGITAL PATTERN CUTTING: LECTRA

Since its establishment in 1973, Lectra quickly became known as a major player in the arena of CAD for clothing and has powered through the decades by effectively responding to changing industry challenges and is currently the world leader in the field of digital pattern cutting technology. Recent acquisitions of key businesses such as Kubix Lab, Retviews, Neteven, Gerber, and Gemini CAD have assisted Lectra to reinforce its position as an Industry 4.0 player, and this in turn has enabled Lectra to effectively automate digital pattern cutting processes required for apparel product development and apparel production development. Lectra's software offers currently include Retviews, Neteven, Kaledo, Modaris, Quick Offer, Flex Offer, Kubix Link PLM, PIM, and DAM, which cumulatively support the analysis of competition, publication of product to marketplaces, design, development, and Product Lifecycle Management (PLM) of apparel whilst Vector and Fashion-on-Demand, including Virga cutting solution, support production planning and manufacturing. The key features and latest advances of some of these tools will be outlined in the rest of this section.

7.5.1 Lectra Retviews

Although Lectra Retviews is not directly involved with digital pattern cutting, it is an important tool that essentially assists in decision-making, using automated AI-based digital technologies to compare fashion competitors by analysing product categories and prices. This in turn informs the development of products that meet customer expectations and is therefore highly effective where on-demand approaches are concerned.

7.5.2 KALEDO

Analysis from Lectra's Retviews fashion marketing analytical software can be utilised by fashion companies to make merchandising decisions when developing new products and collections. Fashion and textile designers will create initial stylised drawings and storyboards using Lectra's Kaledo Style and Kaledo Textile solutions. These applications are extremely sophisticated, particularly where elaborate woven and knit coordinates and colourways are concerned. Kaledo Textile applications for print, weave, and knit design enable textile designers to create new textile designs in true or indexed colour or to modify print repeats interactively. Colour matching can be aligned with Pantone, Munsell, NCS, Coloro, and CSI colour libraries, and it is also possible to communicate spectral colour values, for example, to accelerate lab dip creation.

The Kaledo Print application also has a draping studio that can create realistic 2D-mapped simulations for presentation purposes. In terms of textile design, stripes can be created freehand or with precise technical tools. It is also possible to modify yarns and weave patterns and adjust weave density interactively by selecting from an inbuilt 3D yarn studio. Also, it facilitates the accurate design of knit products using 3D stitches and yarns via a dynamic 3D yarn studio. Accurate technical drawings and specifications can be created with Kaledo Style as the vector-based drawing tools are supported by a library of customisable lines and stitches. Overall, this software enables apparel and textile designers to precisely communicate their designs, ensuring design integrity remains intact, thus avoiding errors due to misinterpretation of design-based data.

7.5.3 MODARIS

Lectra's Modaris digital pattern cutting software was originally introduced in 1996 and through concurrent development and innovation is still currently one of the leading digital pattern cutting systems today. The Modaris portfolio which links with cloud-based fabric consumption (Quick Estimate), Kubix Link PLM and Made to Measure solutions, include versions to suit different business models, including Modaris Essential, Modaris Classic, Modaris 3D, Modaris Expert 2D, and Modaris Expert 3D.

7.5.4 MODARIS EXPERT 2D

This application is used in situations where lean product development is required, particularly for complex pattern styles which have extensive components or size ranges or those which have highly complex constraints such as exacting measurement tolerances. In essence, this software automates low value-added tasks and enables pattern cutters to spend more time cutting patterns rather than engaging with time-consuming workflows.

A key feature and advancement within Modaris Expert is the ability to partially automate the pattern alteration process which, depending on the complexity of the pattern, can be extremely complicated and time-consuming and can result in

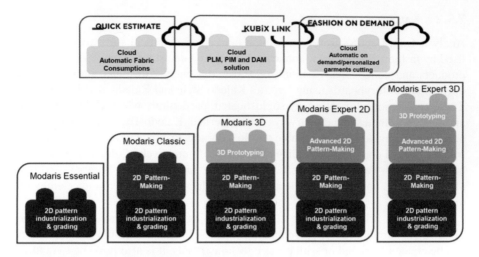

FIGURE 7.2   Lectra Modaris Solutions portfolio 2023.

Source: Courtesy of www.Lectra.com

discrepancies as each component may need to be individually altered. Automated pattern alterations are enabled through the creation of links between pattern pieces called dependencies. The higher the number of pattern pieces and sizes, the more time is saved. So, for example, if you needed to alter the back yoke seam on a pattern for a lined coat that was created with dependencies, you would need to complete just a few initial changes, and the software can then complete the rest automatically. Without dependencies, you would need to alter every associated component including the lining; so this application saves time and costly mistakes.

Modaris Expert 2D also enables fast-track geometric constructions because parallels, rotations, and linked points are all tied with dependencies. This means you need to change only the reference geometry to automatically update all dependent geometries across all pattern pieces and sizes. There is also an automatic-pleat update tool, which enables the automatic adjustment of pleat shape, pleat number, and depth that can be applied to each garment size at any time in the pattern development or grading process. This is supported by the introduction of a clever instant pleat opening feature.

A new dart construction rotation tool within Modaris Expert 2D enables the instant opening, closing, and sharing of darts, and initial model modifications enable all derivative designs to be updated accordingly.

7.5.5   Modaris Expert 3D

Modaris Expert 3D is supported by a virtual prototyping tool equipped with powerful 3D simulation and collaboration tools. This combination vastly speeds up the design and product development process and the decision-making process because changes to style or fit that can ordinarily require repetitive sampling are eliminated, and this in turn reduces physical prototyping. Virtual prototypes can be digitally shared using 360-degree videos, accessible on any device, meaning designers and buyers can visualise, comment, and approve the style and fit using an integrated 3D style module. The cost of product development is therefore reduced as is the overall time to market.

The latest advancements and updates offered within Modaris 3D include a variety of new additions to the asset library including a wider variety of fabrics. It is also compatible with high-quality fabric scanners such as Vizoo xTex scanner. This is an important development, as use of accurate fabric data during the 3D design process can ultimately make the digital renderings of the prototypes more true-to-life and accurate. Other asset additions include 3D topstitching effects, realistic scenes, lighting studios, as well as a Pantone® and Natural Color System®. Another recent advancement within Modaris (V8R2) is a special dart feature to help pattern makers add dimension to their garments with ease. This new dart feature also makes the process of modifying a dart much quicker.

Unlike many other digital pattern cutting systems, Modaris can manage different units of measurement, and this is highly beneficial for brands that need to guarantee size compliance especially where different measurement systems are in place.

7.5.6   QUICK OFFER

Lectra has recently introduced an innovative solution known as Quick Offer for fabric cost estimation and procurement needs. The relevance of this software solution for American, English, and European brands and retailers can be contextualised by the fact that 70% of them prepare markers for their suppliers to control fabric costs and waste management. The number of suppliers who control marker making is likely to increase further in the future as brands and retailers search for competitive solutions related to the increasing cost of raw materials and sustainable design, development, and manufacturing practice.

Quick Offer encompasses two cloud-based applications called Quick Estimate and Quick Nest as part of its Industry 4.0 strategy. These applications assist in automating fabric consumption calculations during the product development and production development processes. Quick Estimate allows pattern makers to estimate fabric consumption during the product development process whilst working within Modaris whilst Quick Nest, which is accessed from Lectra's marker manager application, allows precise fabric quantities to be calculated and fabric orders placed for production by simultaneously processing multiple marker lists.

7.5.7   FLEX OFFER

Launched in 2021, Flex Offer© is a SaaS cloud-based automated nesting solution developed specifically for manufacturers with high-volume nesting. The software utilises powerful algorithms that enable the simultaneous processing of all nesting requests using cloud-based processing, which means that manufacturers are no longer limited by computer processing speed and capacity.

The software is licensed on a subscription basis and assists manufacturers to maintain or elevate their profit margins by controlling fabric utilisation and costings during the marker making process. This is especially relevant considering current industry challenges such as increasing consumer demand for competitively priced new-styles-more-often and made-to-measure apparel. Bearing in mind that fabric can constitute two-thirds of the cost of a garment, enabling automation of the fabric costing and marker making process can help maintain or elevate margins and overall profitability and eliminate other issues such as fabric surplus and fabric shortages whilst also delivering effective solutions to minimise or eliminate fabric wastage. Flex Offer has three activity streams including:

- Flex Nest Cost & Bid focuses on fabric costing and enables suppliers to respond quickly to request for quotes (RFQs) and to verify that fabric costs remain on target whilst meeting the expectations of contractors and buyers with accurate bids that protect margins.
- Flex Nest Procurement focuses on activities related to reducing the cost of fabric orders as well as avoiding excessive inventories. This tool greatly supports the management and control of markers especially during periods of peak activity.
- Flex Nest Production focuses on the preparation of highly efficient markers that protect margins and therefore greatly assists with fabric waste reduction and meeting sustainability goals.

7.6 CONCLUSIONS

This chapter has provided an overview of the key advancements and pioneering Industry 4.0 solutions within the field of digital pattern cutting technology for both product development and production development. It is apparent that Gerber and Lectra are currently the main global players delivering a fully integrated digital pattern cutting system that incorporates 2D and 3D CAD with made-to-measure, marker making, nesting, cutting, and PLM capabilities.

REFERENCES

Barrie, L (2004a) *Lectra Buys Investronica in Euro 51.5 Million Deal* [online], www.just-style.com/news/france-lectra-buys-investronica-in-eur51-5m-deal/?cf-view&cf-closed (Accessed 02/01/2024)

Barrie, L (2004b) *FRANCE: Lectra to Keep Investronica Brand—CEO* [online], www.just-style.com/news/france-lectra-buys-investronica-in-eur51-5m-deal/?cf-view (Accessed 02/01/2024)

Baytarm, F and Sanders, E (2020) Computer-Aided Patternmaking. In: Moore, G (ed.) *Pattern Making History and Theory*, Bloomsbury, London, pp. 109–126

JustStyle (2018) *Digitalisation Key to Meeting "See Now Buy Now" Shift* [online], www.just-style.com/news/digitalisation-key-to-meeting-see-now-buy-now-shift/ (Accessed 02/01/2024)

Lectra (2021) *Fashion Market: Meet Demands with Style* [online], www.lectra.com/en/fashion (Accessed 02/01/2024)

Wright, B (2021) *Lectra Completes Gerber Technology Purchase* [online], www.just-style.com/news/lectra-completes-gerber-technology-purchase/?cf-view# (Accessed 02/01/2024)

8 Recent Advances in 3D Fashion Design Technology

8.1 INTRODUCTION

This chapter will open by providing an overview of the development of 3D fashion design software. It will move on to consider the current and emerging drivers of this technology, including business confidence in 3D fashion design software, cost of sampling versus cost of computing technology, sustainable design practice, education, the global pandemic, consumer demand for faster fashion and personalised apparel, trends in the use of virtual try-on using augmented reality to support e-commerce, the growth of new digital fashion markets, and the expanding Chinese fashion industry. The chapter will also provide details of recent and emerging advancements in 3D fashion design software through the lens of a new 3D fashion design software company Style3D and finally a consideration of the future of 3D fashion design software. To provide the most current insight into advances in 3D fashion design software, key contributors include Eric Liu, Founder and CEO, Style3D; Alfie Chen, Chief Operating Officer and Chief Marketing Officer, Style3D; and Rebecca Ma, Digital Marketing Director, Style3D,

8.2 DEVELOPMENT OF 3D FASHION DESIGN TECHNOLOGY

One of the first recorded 3D applications for cloth simulation emerged in the early 1990s. It had basic simulation capabilities, consisting of a cloth sample over a geometrical shape (Song and Ashdown, 2015). Despite the early emergence of 3D simulation technologies, demand for and acceptance of 3D fashion design software within the fashion industry has lagged other key industries.

According to Makryniotis (2015), the fashion industry is a late adopter of technology and although it relies on innovation due to its size, and for socio-economic reasons, new technologies have taken time to spread and establish as industry standards or common practices (Makryniotis, 2015, p. 6).

Most of the well-known 3D fashion design software companies that we know today such as CLO3D, Browzwear, Optitex, and Tukatech were established between the 1990s and early millennium period (Table 8.1) at a time that coincided with intensified globalisation of the fashion industry and the introduction of fast fashion for which this software offered tangible solutions to faster sampling, reduced waste associated with sampling, and a reduced carbon footprint associated with the transit of samples from East to West. Companies such as Style3D have recently entered the market in response to the rapidly expanding Chinese fashion industry.

DOI: 10.1201/9781003126454-8

TABLE 8.1

Timeline of the Establishment of Key 3D Fashion Design Software Companies

Year Est.	Software	Company Name	HQ Location
1995	Tuka3D	Tukatech	Los Angeles, United States
1999	V Stitcher	Browzwear	Singapore
1995	Optitex	FOG Software Group	Israel
2007	Modaris 3D Fit	Lectra	France
2009	CLO3D		Seoul, Korea
2014	AccuMark 3D	Gerber	New York, United States
2015	Style3D	Zhejiang Linctex Digital Technology	Hangzhou Shejiang

8.3 KEY DRIVERS OF ADVANCEMENTS IN 3D FASHION DESIGN SOFTWARE

The following list provides a summary of some of the key drivers that are supporting advancements in 3D fashion design software, each of which will be discussed in more detail in the rest of this section:

- Business confidence in 3D fashion design software
- Cost of sampling and cost of computing technology
- Sustainable design practice
- Education
- The global pandemic
- Consumer demand for faster fashion, smaller ranges more frequently and made to measure
- The e-commerce industry for virtual try-on using augmented reality
- Digital fashion
- Chinese fashion industry now the largest with rapid ongoing expansion

8.3.1 BUSINESS CONFIDENCE IN 3D FASHION DESIGN SOFTWARE

Early editions of 3D fashion design software lacked photorealistic qualities, therefore fashion business leaders lacked confidence in its capabilities to represent accurate fit and form, believing the 3D renders did not look realistic and may therefore not provide accurate information and pose a risk in progressing the sample sealing process (Guest, 2021). This position has changed in recent years as photorealistic qualities have dramatically improved in line with improvements in computing technologies (Guest, 2021).

8.3.2 COST OF SAMPLING AND COMPUTING TECHNOLOGY

The cost of physical sample making has been a key driver in determining the uptake of 3D fashion design technology. For example, the cost of sample making in the fashion industry has always been minimal at around a few hundred pounds per sample, making the case for switching to 3D fashion design software much harder. This is in direct contrast to the automotive industry where physical sampling starts at around US\$1 billion per sample car, making the case for physical sampling prohibitive and the switch to 3D digital sampling unquestionable. As labour costs in the fashion industry continue to rise, the case for switching to 3D fashion design software for sample prototyping is becoming more necessary, specifically during the early-to-mid stages of product development where the physical sampling of several prototype garments can be eliminated.

Another element of cost that is assisting to drive this technology is the cost reduction in the computer hardware needed to run the 3D software. Until recently, the cost of computers that met the minimum operating specification was cost-prohibitive for many businesses or individuals. Since 2020, most of the entry-level gaming laptops are now affordable and have fast-enough graphic cards to run the software (Comen, 2021). Likewise, the reduction in software cost through the introduction of SaaS models has collectively meant that 3D fashion design software is now accessible to anyone who has a standard gaming computer and can pay a small monthly subscription fee. The combination of these developments has meant that hardware and software have become advanced enough and cheap enough to support more widespread uptake. Many of the established 3D fashion design software companies now also offer a cloud-based rendering service, and some have partnered with render farms such as CLO3D and iRender GPU, a cloud-based rendering service providing an alternative to upgrading or replacing personal computers. Rendering costs are generally inexpensive and start at around a few pounds per hour depending on the complexity of the render.

8.3.3   Sustainable Design Practice

Influential designers and wealthy brands have also acted as catalysts to the adoption of 3D fashion design software for purposes of sustainability. For example, in reaction to the ongoing waste and pollution produced by the fashion industry, designers and brands such as Katherine Hamnett and M&S in the late 1990s and early millennium period began to question and lobby the social and environmental impact of the fashion industry (Jacobs, 2020). In 2004, M&S in collaboration with a key supplier, Dewhirst, successfully trialled 3D fashion design software for a range of products using Optitex 3D fashion design software. These events helped to pave the way for the concurrent development and acceptance of 3D fashion design technologies in promoting the wide-ranging benefits that included faster prototyping, a reduction in the volume of samples being moved around the world, a reduction in the volume of physical samples being, and therefore the subsequent reduction in textile waste and effluents involved with dyeing and wet processing of apparel.

8.3.4   Education

Another long-standing barrier of 3D fashion design software uptake has been the labour market because there is currently a limited talent base to recruit from, partly fuelled by long-standing issues relating to software cost and limitation in computing technology but also because the fashion industry has traditionally not taken on board the need to invest in new technology and to upskill/educate its workforce and has long relied on poaching talent from other brands or businesses (Honeyman, 2000). This poaching approach is not currently possible as the pool of talent is still quite empty, made up of a select group of specialist 3D design freelancers, many of whom are self-taught or have transferred from related industries, such as gaming or film, which use similar 3D software, animation, or modelling software such as Marvelous Designer, Autodesk Maya, Autodesk 3ds Max, Rhino 3D, or Foundry's Modo (Roberts-Islam, 2020). However, this approach is changing, and companies like Levis Strauss have established in-house training programmes to train their staff for a digital future, which they claim improves staff retention and minimises the likelihood of employees moving to competitors once trained (Levis, 2021).

Until recently, the higher education sector has also been sluggish in supporting the development of digital fashion skills such as 3D fashion design software. There are exceptions such as London College of Fashion and specifically Manchester Metropolitan University who have pioneered the teaching of 3D fashion design software since the 1990s, using Browzwear and later Optitex, CLO3D, Gerber 3D, and Lectra Modaris 3D software. Many other fashion schools and universities have begun to integrate the teaching of 3D fashion design into their fashion design curriculum, and so graduates with good 3D skills are starting to exit into the employment arena, but more needs to be accomplished by the higher education system to ensure it can meet the requirements of an increasingly digitalised industry. A recent government paper, the Augar Report, has further highlighted the role that higher education needs to play in reversing the issue of digital literacy in the UK and concluded that the UK as a nation of nine million people is 'lacking in strong digital literacy' (Hughes, 2019, p. 26).

8.3.5 THE GLOBAL PANDEMIC

Probably, the most recent and most influential driver of 3D fashion design software has been the recent global pandemic which commenced in March 2020. The pandemic highlighted the fragility and fragmented nature of the fashion industry, which is predominantly reliant on physical as opposed to digital working practices. This lack of connectivity further emphasised the need to create more efficient and sustainable approaches of working (Kalypso, 2020).

Bearing in mind, many of the original 3D fashion design software companies were established up to 30 years ago (Table 8.1) contrary to much of the current media hype that prior to the recent global pandemic only a minority of fashion industry businesses had adopted 3D fashion design software which often included large sportswear, lingerie, and luxury brands such as Nike, Adidas, and the PVH group which includes Tommy Hilfiger, Calvin Klein, Warners, Olga, Kenneth Cole, Michael Kors, and True & Co. Uptake of e technology has accelerated both during and post-pandemic and is likely to accelerate further as more fashion design graduates enter the employment field because transitioning to this technology is often slower and more challenging for mid-career senior designers who originally trained using traditional fashion school methods for three reasons. First, because upskilling in 3D takes time, and most professionals are often pressed for time in an industry that already demands long hours that often stretch into evenings and weekends.

Second, current design practice within the fashion industry often aligns with existing employers' established business processes and structures; therefore, in the short term, the need for mid-career professionals to work with digital fashion technologies becomes less pressing (Roberts-Islam, 2020). And third, until recently, most current 3D software were designed for other industries such as the film, gaming, automotive, or aerospace industries and were arguably not fully adapted to suit fashion design practice. According to Gettini (2020, cited in Roberts-Islam, 2020), even now, most are driven by technical specs and are pattern dependent, meaning designers need to fundamentally change their creative practice in favour of a keyboard and mouse. Gettini believes the solution for the future may be to 'expand the designers tool kit' and create a system that is developed with fashion designers in mind so that they can still use a tablet and pen that automatically digitises what they draw.

Therefore, investment for lifelong learning, continuous professional development, as well as developing 3D fashion design software platforms that embrace traditional design practice are currently considered essential ingredients for further successful adoption of 3D fashion design software that will ultimately support enhanced sustainable design practice.

8.3.6 Consumer Demand for Faster Fashion and Made to Measure

The fashion industry is currently facing several key challenges, some of which are driven by consumer demand which include faster fashion as well as the trend towards personalised apparel. Despite the fact that the consumer has been active in pushing for sustainable fashion practice, fast fashion continues to flourish. Trends are constantly changing at a rapid pace, as consumers increasingly adopt a 'see now buy now' mentality (JustStyle, 2018). Therefore, brands need to produce a larger number of small-volume collections with a greater variety of style options that are competitively priced and delivered within a shorter time frame to remain competitive. At the same time, changes in diet, lifestyle, and demographics have meant that many brands are producing clothes using fit parameters that have become out of date, and this has meant high return rates and elevated costs associated with the administration of returns. In response to this, and to meet the growing trend towards personalised apparel, some brands have diversified their business models to offer made-to-measure apparel produced using on-demand manufacturing methods (Lectra, 2021). 3D fashion design software provides an effective solution in drastically reducing the product development time and waste involved in all of these scenarios. Instead of it taking a day to make a physical sample, several samples can be digitally created, evaluated, and perfected within the same time span with no physical waste.

8.3.7 The E-Commerce Industry for Virtual Try-On Using Augmented Reality

Virtual apparel try-on is a technology that has existed for several decades and in its simplest form enables digital clothing to appear on a person as they look into a smart mirror or other web-enabled device that uses augmented reality. According to McDowell (2021), research and development into virtual try-ons using AR has accelerated post-pandemic with investment from both start-ups and major tech companies, such as Snapchat parent company Snap Inc, because it is believed this technology has the potential to unlock digital clothing sales and help to reduce e-commerce returns. Virtual try-ons are totally dependent on 3D fashion design software which forms the first step in creating a virtual try-on digital asset.

8.3.8 Digital Fashion

The first digital-only NFT dress called the 'Iridescence Dress' which was designed using 3D fashion design software sold for US$9,500 in 2019 (The Fabricant, 2019). According to BOF (2021), digital demand for fashion and luxury brands is forecasted to increase to over six billion dollars in the next five years representing an entirely new revenue opportunity with gaming and social media being the two key areas for individuals to showcase their digital fashion assets. However, none of this is possible without 3D fashion design software which is the essential ingredient for designing and creating digital fashion for the metaverse.

8.3.9 Chinese Fashion Industry

New and expanding fashion markets such as the Chinese fashion industry, which is now the largest fashion market in the world, worth over two billion dollars (McKinsey, 2019), is supporting the successful establishment of new 3D fashion design software companies such as Style3D, which was launched in 2015.

This is an important technology for the Chinese fashion industry as it enables rapid prototyping and seamless communication of design and product development data from design studios in China to factories that have been migrated to countries such as Vietnam, Cambodia, Bangladesh, and Ethiopia because of rising labour costs in China.

8.4 FOCUS ON STYLE3D FASHION DESIGN SOFTWARE AND DIGITAL FABRICS

Style3D fashion design software offers similar tools and features as other key competitors operating in the same 3D fashion design software market. For example, it has a familiar split-screen user interface with a 2D window displaying 2D pattern development and a parallel window displaying a 3D avatar/3D garment simulation. As standard, a virtual library is accessed via a drop-down menu, offering a variety of avatars, fabrics, hardware/trims, and garments. There is also an import function to enable new assets such as avatars created from 3D body scanning or from companies

FIGURE 8.1 Style3D Red Multi Puffa Jacket.

Source: Courtesy of Style3D https://linctex.com

producing photorealistic avatars such as Daz3D to be imported. Existing avatars can be adjusted within the software to create a digital twin of a real body for mass customisation or made-to-measure purposes.

Similar to other leading 3D fashion design software, Style3D has a variety of menus that enable specific design features to be created, such as elasticating, shirring, taping, the addition of interlining, shoulder pads, or changing fabric grain direction. Other common tools enable the quilted appearance on outerwear or the simulation of authentic finishes on casual or denim products. Drop-down menus are also available

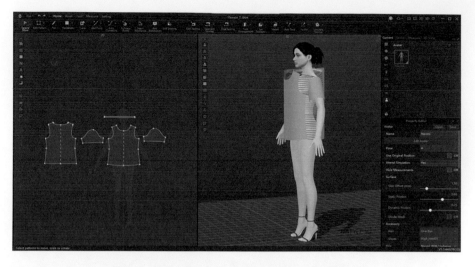

FIGURE 8.2 Style3D fashion design software: main user interface.

Source: Courtesy of Style3D https://linctex.com

for viewing and editing items such as fabric, buttons, zips, topstitch, and stitch type or to introduce and edit the degree of seam pucker, which is used to create a realistic sewn appearance. Similar to other leading 3D fashion design software companies, Style3D Tech Packs are automatically generated, and this assists in eliminating mistakes that are often present in analogue-based tech packs.

A feature that sets Style3D apart from other leaders in this market is that Style3D comprises a unique set of interrelated tools that connect the Style3D fashion design software with a light touch PLM system to support communication and a specialised fabric testing system that enables the physical testing and scanning of fabric for the generation of digital materials. Most other companies do not offer this combined package and instead invest in separate tools. For example, there are several highly specialised fabric scanners in the market including the Vizoo xTec fabric scanner that uses a combination of scanning technology and powerful software to generate high-quality fabric scans that support photorealistic 3D visualisation.

Even though the latest fabric scanners can scan fabrics in just a few minutes, generating digital materials can be time-consuming as the physical testing can involve measuring parameters such as weight, thickness, drape, bendability, as well as stretch in the warp, weft, and bias direction. This is a current limitation for companies that are designing for fast fashion or on-demand methods which can involve many different textiles. In response to this limitation, Style3D is working to develop alternative approaches using AI and machine learning to automatically generate fabric properties that will take minutes rather than hours to generate whilst delivering more precise and realistic results.

A related new development that may support the automated creation of digital materials is the development of U3M, created by Vizoo and Browzwear in 2018.

U3M is the first digital material format in the industry to combine digital material formats together in a consistent manner, thus enabling industry standardisation of digital material libraries (Browzwear, 2018).

Just like its competitors, Style3D works closely with its users and their feedback to support concurrent software improvements and developments, which culminate in regular software updates. A recent and impressive advancement by Style3D is the ability to virtually tighten a waist corset by supressing the virtual waist, thus simulating the fleshy nature of human body which has not before been successfully simulated. Figure 8.3 shows the drawcord being progressively tightened.

Another recent advancement by Style3D is the ability to stabilise collision using AI. So, for example, if the 3D designer does not set the initial position correctly, the Style3D software can quickly establish the correct shape of the design and prevent collision from occurring. The following set of four pictures demonstrates this anti-collision feature using a yank of cloth draped over metal bars at either.

8.5 FUTURE OF 3D FASHION DESIGN SOFTWARE

A long-standing limitation of 3D fashion design software has been the complexities associated with simulating the deformable soft fleshy human body, and although most of the leading 3D fashion design software platforms support fit evaluation using virtual stress maps along with a variety of virtual measurement and articulation

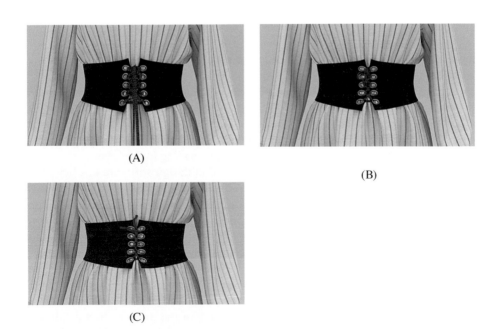

(A)

(B)

(C)

FIGURE 8.3 Style3D virtual suppression of waist.

Source: Courtesy of Style3D https://linctex.com

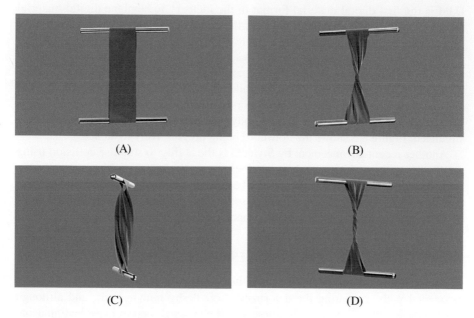

(A)   (B)

(C)   (D)

FIGURE 8.4   Style3D anti-collision technology.

Source: Courtesy of Style3D https://linctex.com

tools, physical prototyping is still needed to accurately confirm that the prototype actually fits a real human body.

In response to this, Style3D is in the process of creating an avatar that simulates the soft and deformable features of the human body which can vastly improve fit assessment. This potential advancement could drastically further reduce the volume of samples and the subsequent waste associated with the physical prototyping process and may even assist to end all physical sample making in the future.

Another long-standing limitation of 3D fashion design software has been the lack of 3D virtual fasteners and trimmings that are also commercially available as physical products that can be procured in the physical world. This situation is progressively improving as companies, such as YKK—global leaders in fasteners and trims, have developed partnerships with leading 3D fashion design software companies such as Browzwear. This advancement enables designers to seamlessly incorporate 3D models of YKK products into their design and virtual tech packs making the product development process faster and error-free (Browzwear, 2019).

There have also been important advancements in the area of digital fabrics. As outlined earlier in this chapter, digital fabrics can be created by combining digital scan data with physical fabric testing data. However, a limitation with digital materials is that unless you have a physical fabric swatch there remains uncertainty about the physical handle of the fabric which is a key design and comfort consideration. Over the last 20 years, there has been ongoing research and development in Haptic Gloves for the purpose of simulating tactile sensations of virtual objects (Perret and Vander Poorten, 2018; Chan and Torah, 2021, Kim et al., 2021). The most recent and successful haptic glove currently on the market is the HaptX G1.

HaptX G1 gloves enable collaborative workflows whereby users can feel the same objects, regardless of the users' physical location. Inside the gloves are hundreds of microfluidic actuators that physically displace your skin, so when you touch and interact with virtual objects, the objects feel real. A flexible, integrated tendon system applies up to 40 lbs of resistive force per hand, so you feel the size and shape of virtual objects, and a magnetic motion capture system tracks 30 degrees of freedom per hand with sub-millimetre precision with apparently no perceivable latency. This glove has been developed to enhance immersive experiences for users in the metaverse and is sensitive enough to simulate the feeling of raindrops. The company is working on other new developments with even greater levels of sensitivity, and this provides an encouraging step towards the creation of a haptic glove dedicated to simulating fabric handle which will be a valuable tool for supporting digital product development of fashion in the future.

In the same way that 3D fashion design software companies are partnering with key fastening and trimming companies, their partnership with fabric mills is also a key requirement for the future. There is already evidence to show that these relationships are already developing, and it is interesting to see that the knitwear sector is embracing this. For example, market leader Shima has recently released Apex Fizz 3D fashion design software for the knitwear industry. The software enables knitwear designers to select from over 1,500 knit patterns whilst simulations of fabrics stored within Shima's virtual yarn bank can replicate the textures of yarns from industry-wide spinners. Data for all yarns held within the yarn bank can be downloaded and used within Apex Fiz and then simulated into a fabric/design.

Perhaps, in the near future fabric selection at key events such as Premiere Vision might take the form of an immersive design experience using, for example, a VR platform such as Spatial where designers can meet with fabric agents in a portal, and both see fabric and feel fabric ranges with oculus and haptic gloves whilst designing and fitting in 3D.

The most successful 3D fashion design software systems of the future will be those that offer a highly accurate and seamless collaborative workflow that spans design, product development, manufacture, and retail. A system that allows designers to maintain their original design practice and a sustainable system that enables the accurate evaluation of fit and comfort so that manufacturers can commence production without the need for physical sampling.

FIGURE 8.5 HaptX G1 gloves to simulate realistic touch for the metaverse.

Source: Courtesy of HaptX www.Haptx.com

REFERENCES

BOF (2021) *Metaverse: A $50 Billion Revenue Opportunity for Luxury* [online], www.businessoffashion.com/news/technology/metaverse-a-50-billion-revenue-opportunity-for-luxury/ (Accessed 02/01/2024)

Browzwear (2018) *U3M Driving the Fashion Industry's Digital Transformation through Consistent and Unified Data* [online], www.just-style.com/news/browzwear-and-vizoo-standardise-digital-materials/ (Accessed 02/01/2023)

Browzwear (2019) *YKK Partners with Browswear* [online], https://browzwear.com/blog/ykk-partners-with-browzwear (Accessed 02/01/2024)

Chan, J and Torah, R (2021) *E-Textile Haptic Feedback Gloves for Virtual and Augmented Reality*. Presented at the 3rd International Conference on the Challenges, Opportunities, Innovations and Applications in Electronic Textiles (E-Textiles 2021), Manchester, UK, 3–4 November [online], www.mdpi.com/2673-4591/15/1/1 (Accessed 02/01/2024)

Comen, E (2021) *USA Today Tech* [online], https://247wallst.com/special-report/2018/06/18/cost-of-a-computer-the-year-you-were-born-2/ (Accessed 02/01/2024)

The Fabricant (2019) *Iridescence* [online], www.thefabricant.com/iridescence (Accessed 02/01/2024)

Guest, SF (2021) So What's Next for 3D in Fashion . . . And What Brands Must Demand from the Technology. *Sourcing Journal* [online], https://sourcingjournal.com/topics/technology/style3d-3d-zhejiang-design-studio-fashion-cloud-visualization-metaverse-317323/ (Accessed 02/01/2024)

Honeyman, K (2000) *Well Suited: A History of the Leeds Clothing Industry*, Pasold Research Fund, Oxford

Hughes, D (2019) *The Post-18 Education Review, The Augar Review* [online], https://commonslibrary.parliament.uk/research-briefings/cbp-8577/ (Accessed 02/01/2024)

Jacobs, B (2020) *Katherine Hamnett: The Original Fashion Eco Warrior* [online], www.bbc.com/culture/article/20200113-katharine-hamnett-the-original-fashion-eco-warrior (Accessed 02/01/2024)

JustStyle (2018) *Digitalisation Key to Meeting "See Now Buy Now" Shift* [online], www.just-style.com/news/digitalisation-key-to-meeting-see-now-buy-now-shift/ (Accessed 02/01/2023)

Kalypso (2020) *The 2020 Digital Product Creation Survey Briefing* [online], https://kalypso.com/files/docs/Exec-Summary-Annual-Retail-Innovation-Adoption-Survey-2020_2020-12-17-215126.pdf (Accessed 27/12/2023)

Kim, H, et al. (2021) Recent Advances in Wearable Sensors and Integrated Functional Devices for Virtual and Augmented Reality Applications. *Advanced Functional Materials*, 31(39). https://doi.org/10.1002/adfm.202005692 (Accessed 02/01/2024)

Lectra (2021) *Building Customer Intimacy Through On-demand Production in Fashion* [online], https://appsource.microsoft.com/en-us/product/web-apps/lectra1595328015994.fashion-on-demand-2021 (Accessed 29/12/2023)

Levis (2021) *Training Our Employees for a Digital Future* [online], www.levistrauss.com/2021/05/17/machine-learning-bootcamp/ (Accessed 02/01/2024)

Makryniotis, T (2015) *3D Fashion Design*, Batsford, London

McDowell, M (2021) *Why AR Clothing Try-on Is Nearly Here* [online], www.voguebusiness.com/technology/why-ar-clothing-try-on-is-nearly-here (Accessed 02/01/2024)

McKinsey (2019) *The State of Fashion* [online], www.mckinsey.com/~/media/mckinsey/industries/retail/our%20insights/the%20state%20of%20fashion%202019%20a%20year%20of%20awakening/the-state-of-fashion-2019-final.ashx (Accessed 02/01/2023)

Perret, J and Vander Poorten, E (2018) *Touching Virtual Reality: A Review of Haptic Gloves*. Proceedings of the ACTUATOR 18, Bremen, Germany, 25–27 June [online], www.researchgate.net/publication/324562855_Touching_Virtual_Reality_a_Review_of_Haptic_Gloves (Accessed 02/01/2024)

Roberts-Islam, B (2020) Is Digitization the Saviour of the Fashion Industry. *Forbes.com* [online], www.forbes.com/sites/brookerobertsislam/2020/01/07/is-digitisation-the-saviour-of-the-fashion-industry-i-ask-a-cto-who-knows/?sh=242afe457e7a (Accessed 29/12/2023)

Song, HK and Ashdown, SP (2015) *Investigation of the Validity of 3-D Virtual Fitting for Pants* [online], www.researchgate.net/publication/281234726_Investigation_of_the_Validity_of_3-D_Virtual_Fitting_for_Pants (Accessed 02/01/2024)

9 The Gap between Digital and Physical 3D Prototyping for Fashion

9.1 INTRODUCTION

The fashion industry's progression towards full digitalisation of the design and product development process will be partly dependent on achieving a mirroring of the physical and digital apparel prototyping processes. This chapter will outline some of the key differences that currently exist between physical and digital apparel prototyping and how designers and software developers can assist in closing some of these gaps to further enhance the technical accuracy of 3D apparel prototyping. Rather than focus on a specific named software, generic terms which are relevant to the majority of 3D fashion prototyping software will be used so as not to prioritise one specific software over another. The key prototyping processes that will be explored will include stitching, seam pucker, puffer effect, seams, fitting, and fabric performance. These processes were chosen on the merit of their value in the sampling process as well as the existence of a clear virtual representation of each tool in current versions of popular 3D visualisation software.

To provide the most current insights regarding recent and emerging advancements in this area, the key contributor is Kate Ryabchykova, 3D fashion prototyping artist and lecturer at Manchester Fashion Institute at Manchester Metropolitan University. Other contributors include Professor Mykola Riabchykov, specialist in sewing equipment and technologies at the Lutsk National Technical University, Ukraine; Vernita Sothirajah, commercial fashion designer; and Martina Harvey Capdevila, owner of the sustainable brand Original Source and Supply.

9.2 BACKGROUND: THE GAP BETWEEN THE PHYSICAL AND THE DIGITAL IN 3D APPAREL PROTOTYPING

Digital apparel prototyping using specialised 3D prototyping software such as CLO3D, Tuckatech, Browswear V Stitcher, Style3D, Optitex, and many others have come a long way in the last 20 years, but uptake amongst enterprises involved in the design and development of fashion apparel is far from mature. A long-standing barrier to adoption has included lack of confidence in the accuracy of 3D prototyping software, but this position has changed in recent years as photorealistic qualities have dramatically improved in line with improvements in computing technologies (Guest, 2021). Commercial interest and uptake of this technology have also accelerated both during and subsequent to the recent global pandemic when other key benefits such

DOI: 10.1201/9781003126454-9

as the major reduction in the cost and time associated with sample making, along with major improvements in connectivity and collaboration during the design and development stages of sample making have been realised out of necessity (BOF-McKinsey, 2022).

Other key benefits that have assisted to support user uptake within the apparel industry include its ability to effectively facilitate sustainable design practice by reducing the volume of physical samples that are produced as part of a typical product development cycle. This benefit has also had the positive effect of minimising the volume of samples being transited across the globe from East to West for the purpose of style approval because these photorealistic improvements in 3D prototyping software now enable the majority of visual-based sample approvals such as style or colourway change to be agreed by viewing a 3D digital image rather than a physical garment. However, commercial confidence in the *technical* accuracy of 3D prototyping in terms of fit, comfort, and technical construction requires further work as few companies if any will risk progressing from digital prototype straight to manufacturer as there are still a number of areas where the digital prototype does not accurately mirror the physical garment. The rest of this chapter will highlight some of the technical areas of digital prototyping that need further development and refinements if the industry is to further progress towards full digitalisation of the design and product development process.

9.3 JOINING SEAMS: THE GAP BETWEEN THE PHYSICAL AND THE DIGITAL IN 3D APPAREL PROTOTYPING

9.3.1 DIGITAL STITCHING PROCESSES

Sewing tools in 3D prototyping software are considerably simplified up to the point of automatic stitching in some versions and is one of the easiest and most basic tools to comprehend yet is one of the most challenging in terms of its realism.

In most cases, the virtual sewing tool involves a basic set of tools that enable the joining of different configurations of seams by selecting the seam edges that are to be joined. Tools can include a segment sewing tool, tools for free sewing, or sewing tools that allow simultaneous stitching of three or more parts. A recent development in digital sewing tools is the introduction of *automatic sewing* which can save time in the digital joining process by automatically predicting how parts need to be connected to each other. The drawback of most current digital sewing tools is a lack of any choice in the type of seams or even stitch configuration. Details are joined edge to edge, with no seam allowance, thread/stitch strength, or tension specifications and therefore do not mirror the same output achieved when using physical joining methods.

Most of the current software platforms also fail to report on possible sewing damage or fabric tearing such as seam puckering, plucking, and fabric yarn damage. These issues will therefore need to be tested early in the physical sampling process to determine the sewability of the fabric and required sewing/operational settings.

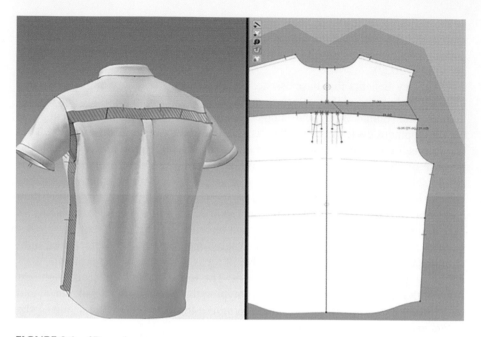

FIGURE 9.1   3D sewing process.

Source: Courtesy of Kate Ryabchykova, www.virtualpandora.com.

9.3.2   PHYSICAL STITCHING PROCESSES

The physical stitching process is much more comprehensive and requires consideration of a wide variety of parameters. The type of fabric required, fabric finishes, technological process, and manufacturing specifications all need to be accounted for to select the appropriate type of needle, thread, and equipment to be used. Sewing skill is also a progressive process that can take years to master, starting at the beginner-level joining of simple superimposed seams using a basic lockstitch sewing machine and progresses to advanced skill levels of seamstresses.

The sewing process for physical fabrics also involves a wide variety of industrial sewing machinery such as the basic lockstitch, overlocker, cover seam or flatlock, and highly specialised cup seamers, seam bonding, and seam welding machines which unlike digital stitching adds a financial component in addition to the required skill set.

Another important element in the physical sewing process is the visual appearance of the stitching within or on top of the seam. Depending on the type of fabric, chosen needle, thread, quality of equipment, and skill level of the seamstress, the stitch may gather or stretch the fabric, create seam puckering, leave marks, and be perfectly stretched or slightly uneven. All these intentional and unintentional outcomes influence the final look of the garment.

9.3.3   THE GAP BETWEEN THE PHYSICAL AND DIGITAL STITCHING PROCESS

The sewing process is therefore one of the major milestones which software developers need to address. Even though the virtual sewing process achieves the basic requirements of joining seams, it fails to account for the variety and complexity of the physical sewing process. The virtual sewing process can be considered to be oversimplified and represents only the basic most commonly known functions of stitching while ignoring decorative sewing, strengthening, seam finishing, quilting, embroidery, or even darning. While it's important to make virtual prototyping more accessible and sustainable than the physical prototyping process, we still need to address the considerable gap between physical and virtual stitching which presently exists.

9.3.4   CLOSING THE GAP BETWEEN THE PHYSICAL AND DIGITAL STITCHING PROCESS

In relation to closing the gap between the physical and digital stitching process, it is important to acknowledge that the level of advanced automation is steadily increasing within the apparel manufacturing sector, marking the movement away from bulk production to the micro-manufacture of small orders and bespoke apparel (McKinsey, 2018; Just-Style, 2020; Postlethwaite, 2022). It is here that digital proto-typing will prove to be a valuable tool in achieving the development of highly accu-rate digital samples that will lead to highly accurate orders, where an order might be as little as one garment, leaving no room for mistakes during the prototyping process. Therefore, closing the gap between the physical and digital stitching process is becoming increasingly important.

One approach towards closing this gap might involve the programming of seam allowance to the patterns and instead of stitching the edge, create internal sewing lines on the pattern. Another short-term solution which may seem counter-intuitive to current research and development into physical 3D sewing on forms, referred to in an earlier chapter, might be to join components on the flat prior to placing them on the avatar, this in turn would enable garment components to be placed on the avatar without reset arrangement. This approach might also in the short term minimise some of the other current challenges commonly experienced during the process of garment placement on a 3D form.

It is becoming increasingly evident that software developers are working hard outputting ever-increasing numbers of small but effective improvements to the digi-tal sewing process such as options to specify stitch tension, as well as intensity and thickness of the stitch. It is anticipated that this pace of change will quickly trans-form virtual sewing to align more closely with current and future real-world sewing processes.

9.4   SEAM PUCKER: THE GAP BETWEEN THE PHYSICAL AND THE DIGITAL IN 3D APPAREL PROTOTYPING

9.4.1   SEAM PUCKER

Seam pucker is considered by the fashion industry to be an undesirable problem to be avoided at all costs as it indicates sewing defects such as tight thread tension or ply shift. In contrast, seam puckering in the virtual world of 3D prototyping is

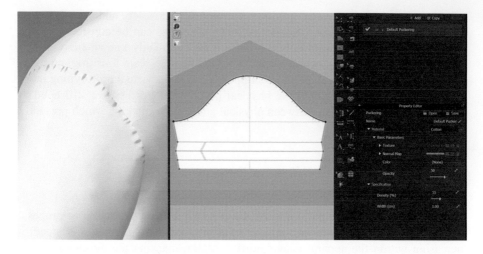

FIGURE 9.2   Seam pucker tool.

Source: Courtesy of Kate Ryabchykova, www.virtualpandora.com.

presented as a stand-alone tool in most 3D visualisation software. Its virtual presence does not indicate any sewing inaccuracies and does not communicate any physical data about the garment at all. Instead, it serves the important function of making digital seams appear more realistic and authentic. This is achieved by digitally adjusting the parameters of the seam pucker tool which involve size, shadow, and highlights.

9.4.2   PRESENCE OF SEAM PUCKER IN PHYSICAL SAMPLES

Seam pucker can be defined as 'an unequal crinkling or a gathering of the seam due to the displacement of the fabric caused by the stitching action' (McLoughlin and Mitchell, 2015, p. 402). There are four main causes of seam pucker that include feeding pucker often caused by the uneven feeding of the top and bottom fabric plies resulting in the shortening of one of the plies, usually the underply. Tension pucker is caused by incorrect thread tension settings as well as a variety of contributing factors, including incorrect needle selection, extension properties of sewing threads, and so on. Stitch density and fabric type are also major causes of seam pucker and are directly linked to thread tension and the length of thread inserted into the seam whereby pucker can result when the stitch density increases the thread tensions within the seam. Inherent pucker results from the displacement of the weave by the needle inserting the thread into the fabric which can become jammed resulting in the puckering of the seam (McLoughlin and Mitchell, 2015).

The American Association of Textile Chemists and Colourists identifies five main levels of seam puckering smoothness. This five-point scale of seam pucker is often used to assess the level of seam pucker that is commercially acceptable with level five being pucker-free (Amerfird, 2010).

9.4.3 The Gap between Physical and Digital Seam Pucker

There is a considerable gap between the virtual and physical representations of seam puckering. Seam pucker in the physical world is an undesirable feature so much so it has its own scientific scale that is used to measure its level of commercial acceptability. In the virtual world, seam pucker is currently considered an essential design tool needed to improve or create the authentic photorealistic qualities of a design.

9.4.4 Closing the Gap between Physical and Digital Seam Pucker

The introduction of the virtual seam pucker tool is a clear example of a successful attempt to bridge virtual versus physical gap. Regardless of what technical specifications dictate, often seam puckering can't be avoided and appears in gatherings, puffer quilts, and other elements, and the closest way currently to bring virtual garments visually closer to real would be to intentionally illustrate 'mistakes' rather than avoid them.

However, this approach might change in the near future as there are new technologies already being used in the 3D printing market that may help to inform how sewing defects such as seam pucker can be eliminated during the prototyping phase. These technologies involve powerful AI algorithms that are programmed to work with 3D software and automated hardware using a generative design process to identify and prescribe alternative ways to resolve design and make problems with the end result being a market-ready product with zero defects (Conn, 2020). So, in the future, the ability to potentially eliminate seam pucker in the physical world may present new challenges for virtual designers and software developers such as the need to develop new ways of representing physical designs in the digital world.

9.5 PRESSURE TOOL: THE GAP BETWEEN THE PHYSICAL AND THE DIGITAL IN 3D APPAREL PROTOTYPING

9.5.1 The Puffer Effect and Insulation Layer in Digital Sampling

The pressure tool is used for a variety of purposes in most 3D apparel prototyping software and is most commonly used to replicate the puffer effect commonly associated with padded coats that are constructed using wadding or feathers. In practical terms, it creates an air bubble between two layers of fabric, a process more commonly associated with the physical construction of floating devices such as life jackets.

To create a puffer effect, two layers of virtual fabric are required. They need to be attached to each other from all sides using a command such as the *layer clone tool*. Internal lines can also be applied and stitched between layers that will give an additional quilt effect. Once the virtual cloth shell has been constructed, opposite values of *virtual pressure* are commonly applied. Virtual pressure volume will determine the strength with which air will virtually press against the fabric, with the option to apply more or less pressure depending on the selected fabric properties and required design outcome. Some 3D prototyping software offers a separate tool designed specifically for the puffer jacket effect. This method is much simpler and only requires

FIGURE 9.3　Puffer jacket designed using 3D fashion software, showing variations in panel design and the use of the seam pucker tool.

Source: Courtesy of Kate Ryabchykova, www.virtualpandora.com

the user to specify which type of filler is required such as wadding or feathers. Users/designers can also specify the weight of the feather and quilting distance.

9.5.2　The Puffer Effect and Insulation Layer in Physical Sampling

In relation to the practical real-world application of creating a puffer effect, air bubbles alone won't cover all the possible scenarios, unless solely referring to inflatable accessories designed for water safety. All garments with puffer effects that are not designed to be placed in the water require as part of their development considerations of comfort and protection and in particular their insulative properties. Such garments would often be worn in mild-to-cold climates during the time of the year when average temperatures would drop below 5–10°C. Thus, the insulation layer which creates the puffer effect will need to be filled with a specific type of material that will be light enough to carry, and at the same time provide enough protection from cold. Goose or duck feathers traditionally were used for these purposes; however, nowadays, the most common material would be synthetic insulation, which

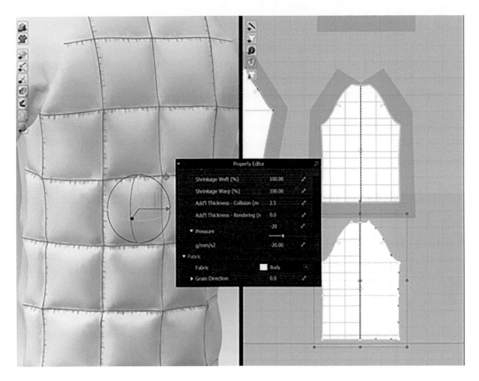

FIGURE 9.4   Pressure tool.

Source: Courtesy of Kate Ryabchykova, www.virtualpandora.com

is made from polyester that has been spun into filaments that create a pocket of air between each fibre. This pocket then warms up from the body temperature and thus provides warmth (Mäkinen, 2009). Whichever filler is used to provide insulation, special pockets on the self and liner fabric will have to be created using rows of stitching, known as quilting, to keep feathers in the same space.

9.5.3   THE PUFFER EFFECT AND INSULATION LAYER: THE GAP BETWEEN THE DIGITAL AND PHYSICAL

The virtual and physical processes of insulation placement are not so different. Both methods would require at least two layers of fabric involving an outer shell and an inner lining that are quilted together to form pockets. However, while a physical puffer effect can be created with a variety of fillers, from natural bird feathers to synthetic filament and air, the virtual sample accounts only for the air-filled. And even though recent software updates by leading 3D prototyping software companies include the possibility to specify details such as feather type and weight, the virtual prototype although aesthetically pleasing to view is still from a technical perspective far from the real sample.

9.5.4 Closing the Gap between the Physical and Digital Puffer Effect

Currently, it is possible to create a high-quality photorealistic sample of a padded or quilted puffer-style jacket using 3D apparel prototyping software; however, translating these features into a physical product is not straightforward. It is indicated that the puffer tool in most current software solutions requires a higher level of sophistication, enabling the designer to select and apply wider range of parameters such as different fillers and quilting designs that can result in a high-quality technical sample that will translate into a physical twin when constructed using traditional methods.

9.6 SEAM TOOL: THE GAP BETWEEN THE PHYSICAL AND THE DIGITAL IN 3D APPAREL PROTOTYPING

9.6.1 Seam Tool for Digital Sampling

Seams in apparel 3D prototyping software are often highly simplified and they differ drastically from the complex technicalities involved in physical seaming.

Seams in 3D prototyping software carry exclusively aesthetic purposes and do not contribute to the physical properties of the garment. There are no specific tools for seam creation, and most of the edge finishing and seaming are performed through the topstitching function with some further adjustments possible to give a more realistic appearance. For instance, commonly used British Standards class three seams (BSI, 1991) involving a bound edge will require a more complex visualisation process than

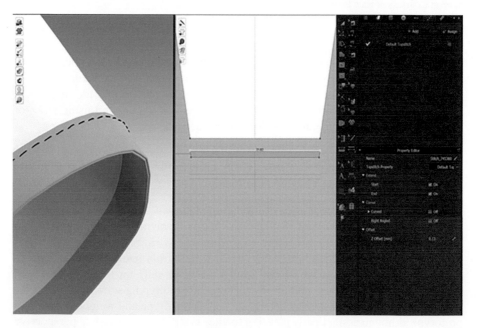

FIGURE 9.5   Seam visualisation tool in 3D fashion design software.

Source: Courtesy of Kate Ryabchykova, www.virtualpandora.com

a British Standards class six seam (BSI, 1991) involving a self-fabric hem turning, whereas it is commonly the opposite for physical seams.

9.6.2   SEAMS USED FOR PHYSICAL SAMPLING

The variety of seams that are used for constructing physical seams are complex and have been categorised by BSI (1991) into eight classifications of seams. Within each class there exists a wide variety of versions that are individually characterised by their use of fabric plies and stitch applied for their construction. A simple summary of each class of seam is provided in Table 9.1, which illustrates the elaborate nature of physical seams in comparison to their representation in the virtual world.

TABLE 9.1
Summary of Seam Classification

Class 1 Superimposed seam(s)	Generally, involve two or more pieces of material superimposed over each other and joined near an edge with one or more rows of stitches. There are various types of seams within this class, and they can be sewn with a variety of stitch types, for example stitch types 301, 401, stitch class 500 (over edge stitch) or combination stitches (e.g., stitch class 516). In terms of application, they are often used to create neat load-bearing seams for lingerie, shirts, etc.
Class 2 Lapped seam(s)	Two or more plies of material are lapped (i.e., with edges overlaid, plain or folded) and joined with one or more rows of stitches such as the lap felled type, involving only one stitching operation which is a strong seam commonly used to protect jeans or similar garments from fraying.
	The lap felled seam is generally sewn with a 401 chainstitch and is most used in jeans manufacture because of its strong construction. Another lap seam is the French seam often constructed using the 301 lockstitch and is commonly used for rain wear and edge stitching front facings on jackets and dresses.
Class 3 Bound seam(s)	Are formed by folding a binding strip over the edge of the plies of material and joining both edges of the binding to the material with one or more rows of stitching. This produces a neat edge on a seam exposed to view or to wear. Stitches commonly used for bound seams include the 401 chainstitch or 301 lockstitch, typically used for finishing the necklines on T-shirts and tops or internal seams on unlined jackets.
Class 4 Flat seam(s)	Sometimes called Butt seams whereby two fabric edges, flat or folded, are brought together and over sewn with stitches.
	The purpose of these seams is to produce a joint where no extra thickness of fabric can be tolerated at the seam, as in underwear or foundation garments. The looper thread(s) must be soft, yet strong, and the cover thread may be decorative as well as strong. This seam is referred to as a flat seam because the edges do not overlap one another, they will be butted together.
	Zigzag lockstitch, chainstitch, or covering stitch (class 600) are often used for the construction of fine knitted garments where seams are required to be free from bulk.

(*Continued*)

TABLE 9.1(CONTINUED)
Summary of Seam Classification

Class 5 Piping seams: Decorative/ornamental stitching	The ornamental stitch is a series of stitches along a straight or curved line or following an ornamental design on a single ply of material. More complex types include various forms of piping, producing a raised line along the fabric surface. In terms of application, the stitching results in decorative surface effects on the fabric, e.g., pin tucks, application of braids etc.
Cass 6 Edge finishing/ neatening	Edge finishing is where the edge of a single ply of material is folded or covered with a stitch. The simplest of these operations is the application of serging in which a cut edge of a single ply is reinforced by over edge stitching to neaten and prevent fraying. This includes other popular methods of producing a neat edge like hemming and blind stitch hemming. Applications include serging trouser panels, flys, facings etc.
Class 7 Attaching of separate items	This seam class involves seams that require the addition of another component onto the edge of a piece of fabric, e.g., elastic braid onto the edge of ladies' briefs. This type of seam requires two components.
Class 8 Single-ply construction	This seam class consists of one piece of fabric that is turned in on both edges. It is most commonly seen in belt loops or belts for which a folder can be attached to the machine. This type of seam requires only one component.

Source: Adapted from Coats (2022)

9.6.3   THE GAP BETWEEN PHYSICAL AND DIGITAL SEAMING

The construction of seams in physical garments is no doubt a complex and technical process, whereas seam construction in 3D prototyping is predominantly a visual element that doesn't require any special knowledge or skills. In terms of some of the most complicated seams such as those involving folds or double folds, it is almost impossible at this stage of software development to virtually replicate these sophisticated constructional details, leaving a gap that appears to have no immediate solution in the virtual garment development process.

9.6.4 CLOSING THE GAP BETWEEN PHYSICAL AND DIGITAL SEAMING

In terms of closing the gap between physical and digital seaming technology, there are several practices that can be adopted to get closer to the sophisticated performance of physical seams. First, it is important to recognise that in some 3D prototyping software, the topstitching tool is often considered a coherent substitute for the seam. As described earlier, seaming is a comprehensive, often multi-step operation, and it needs to be approached it in a similar manner to that used in physical sampling. Second, the development of another seam tool that has the ability for example to fold seam edges, apply wider width to the hem, use double weight on the hem, strengthen involved areas, and apply seam bonding methods might collectively assist in the better presentation and representation of virtual seams.

9.7 FITTING TOOLS: THE GAP BETWEEN THE PHYSICAL AND THE DIGITAL IN 3D APPAREL PROTOTYPING

9.7.1 ASSESSING FIT IN DIGITAL SAMPLING

The sample evaluation process is one of the most fulfilling and realistic stages of virtual garment development. The majority of 3D prototyping software providers have continued to develop fitting tools to make this process as realistic and easy to work with as possible, including its integration with other digital software systems such as Product Lifecycle Management (PLM) software, which can support collaborative fit evaluation by virtually connecting key stakeholders online regardless of location. With this approach there is no need for viewers to have 3D prototyping software knowledge and no requirement to have access to the software within the PLM system.

The most common tools for virtual fit evaluation are garment fit maps, which include stress map, strain map, and pressure points, which can facilitate the assessment of textiles in relation to the body, the assessment of the chosen construction details, and the assessment of the cut of the fabric. In simplest terms, these tools are used to visually identify where there are fit issues with the garment, including identifying where a fabric is being over or under extended in relation to the volume of ease in the pattern. There are other digital tools and processes that can complement the digital fit evaluation process such as the ability to change the pose of the digital fit model/avatar by initiating a dynamic pose such as lifting the arms or creating a knee lift or stride. Changes in pose will be registered by the fit map tools and can therefore assist the user to identify where additional ease may be needed or reduced to improve fit as a consequence of movement. But there remains a missing element of the digital fit evaluation process that is present during real-world fit sessions which is the valuable verbal feedback that human fit models provide when physical fit sessions take place.

9.7.2 ASSESSING FIT IN PHYSICAL SAMPLING

When it comes to the physical sample evaluation, key members of the evaluating team will need to be familiar with the pattern and garment construction techniques, as well as understand the fabric and material properties. Sample production for the physical assessment of aesthetics and fit is one of the most time-consuming, costly, and wasteful processes that also requires the physical attendance of a team of people such as the designer, pattern maker, model, buyer, and so on. The physical fitting process can be lengthy and often involves a minimum of between five and six samples and therefore between five and six meetings before a design can be approved or sealed. Each iteration involves wastage of materials and the congregation of key stakeholders involving travel to and from the fit session, thus unnecessarily increasing carbon emissions, not to mention the transit of each individual sample often

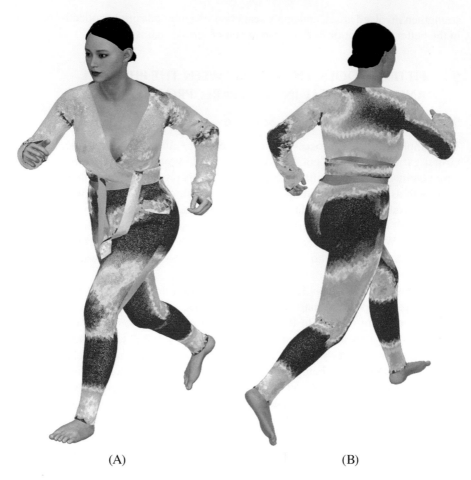

(A)   (B)

FIGURE 9.6   Strain map and pressure fitting maps in 3D visualisation software.

Source: Courtesy of Kate Ryabchykova, www.virtualpandora.com

manufactured in a far-away country. Although the physical sample evaluation process is currently by far the most effective, it is certainly far from being the most efficient and the least sustainable option.

9.7.3   THE GAP BETWEEN PHYSICAL AND DIGITAL FIT ASSESSMENT

In terms of human resources, the virtual and physical sample processes do not differ much. In both cases, the evaluating team should be equipped with sufficient skills and knowledge about garment manufacturing to properly evaluate and fix potential defects. However, virtual sample evaluation offers a range of tools that can assist personnel who are not highly experienced in fit technology to contribute to the process. Drawings, comments, and highlights that are available in PLM systems make it possible for anyone to make an input. On the other hand, virtual samples can often conceal possible issues or discrepancies with garment finishes, seam allowances, hems, and elements of comfort, because some of these features are often simplified in the 3D virtual sample.

9.7.4 Closing the Gap between Physical and Digital Fit Assessment

Even though full adoption and switching to virtual sample development may be an attractive option, it is imperative to plan for at least one physical sample evaluation. As mentioned earlier in the chapter, virtual samples can often conceal possible issues or discrepancies with garment finishes, seam allowances, hems, and elements of comfort that become obvious only when fitted on a physical body. Another key area that is difficult to assess digitally is the internal finish of garments such as seam allowances and finishes which are often overlooked during the early design stage when style and aesthetics such as colour and fit often take precedence. Therefore, greater emphasis on the importance of developing hidden or internal features should become an aspect of good practice for all involved in the design and development of virtual prototypes.

9.8 FABRIC PERFORMANCE: THE GAP BETWEEN THE PHYSICAL AND THE DIGITAL IN 3D APPAREL PROTOTYPING

9.8.1 The Fabric Performance Tool for Digital Sampling

Most 3D apparel prototyping platforms are sold with a comprehensive library of standard fabrics that include denim, cotton poplin, single jersey, fur, cashmere, fleece, voile, and so on.

Each fabric is made available with fabric data to assist the designer to select appropriate fabric for use in an initial prototype. However, for precise prototyping that will be manufactured and retailed, a digital fabric twin of the physical fabric must be created using the combination of a high-quality fabric scanner such as those made by companies like Vizoo along with the physical parameters of the fabric, such as weight, density, modulus, extension, and bendability. These parameters are obtained by using physical textile testing equipment. Some of the 3D apparel prototyping software developers also sell simple fabric testing kits to support the latter. For fabric scanning, some specialised scanning companies offer a fabric scanning bureau service, and this approach is usually taken up by smaller independent companies as some specialised fabric scanning equipment can be cost prohibitive.

9.8.2 Fabric Performance for Physical Sampling

The physical process of the textile evaluation process can be broken into three categories, namely appearance, construction, and function. When it comes to the appearance, in addition to visual and aesthetic parameters of the fabric, this may also include some sensory data such as fabric handle, drape, lustre or dullness, and even smell. Construction will give information as per the type of weave/knit and more specific data such as thread count and yarn construction. Function assessment helps to evaluate textile performance in relation to a specific purpose, in terms of the type

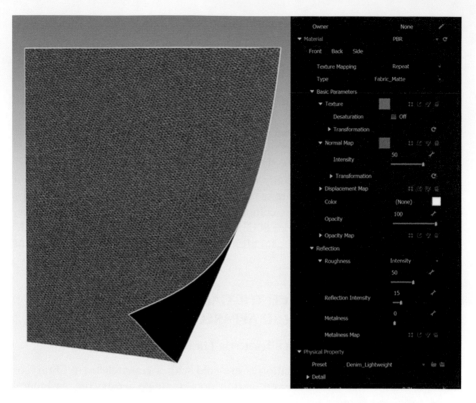

FIGURE 9.7   Fabric sample/properties in 3D fashion design software.

Source: Courtesy of Kate Ryabchykova, www.virtualpandora.com

of the garment or even specific wearer requirements, for example whether textile can cause allergy or next-to-skin comfort etc. At this stage more comprehensive testing may be required, for instance, sample creation, wearer trials, assessment of fit and draping using a physical body for specific products or end use (Ahmad et al., 2017).

9.8.3   THE GAP BETWEEN DIGITAL AND PHYSICAL FABRIC CHARACTERISTICS

In terms of appearance, virtual fabrics achieve quite a high level of realism while in terms of physical performance there are still challenges to be addressed. Ongoing developments in software and hardware technology enable the aesthetic features of virtual fabrics to be realistically and comprehensively represented in terms of colour, print, and even glossiness of the fabric. In contrast, physical fabric values such as stretchiness, tension, strength, yarn density, and weight are substituted with mathematical equations or values and often fail to account for the impact of physical and environmental factors such as weather conditions, humidity, and so on. Instead, virtual fabrics are created using polygon technologies and use universal generic calculations to determine fabric behaviour.

9.8.4   CLOSING THE GAP BETWEEN DIGITAL AND PHYSICAL FABRIC CHARACTERISTICS

While it is now possible to generate digital fabrics that closely represent physical fabrics, and in particular their aesthetic qualities, there are still gaps between digital and physical textiles such as touch, handle, comfort, and smell, which are areas that are being developed by companies such as Koniku's Konikore digital smelling device used to replicate the physical sense of smell.

Other new developments in the area of virtual reality (VR), including VR glasses, which, while still lacking in smell and touch, are beginning to move towards the delivery of a deeper experience of the surroundings and key characteristics of the fabric that, in turn, can trigger the brain to produce related sensory stimuli and recreate quite closely the physical experience of the fabric albeit in a virtual environment (Wang et al., 2021).

9.9   CONCLUSION: RELIABILITY AND ACCURACY OF 3D PROTOTYPING

Most 3D fashion prototyping systems have to date achieved high levels of photorealism; however, there is still a significant gap between the technical aspects involved in the physical and virtual sampling processes. From the areas discussed in this chapter, it is clear that the closest connection is achieved within the fitting process and puffer effect representation. Stitching, seam pucker, seams, and fabric performance still have a long road to go before reaching a realistic application. Even though there is a range of adaptations designers can implement to make virtual samples more reliable, and the fact that software developers are working hard to concurrently release software updates and development that fix many virtual processes, so they more closely mirror reality, there is still a long way to go to bridge the gap between the virtual and physical garment prototyping. Currently, few would risk going straight to manufacturing a digital design that has never been physically sampled at least once, and this level of risk helps to define the level of accuracy and confidence in 3D prototyping software that currently exists in the market. It is anticipated that this remaining gap will begin to rapidly close as industry challenges such as on-demand manufacturing and sustainable design practice continue to push the fashion industry and the associated software developers to further invest in advances that support digitalisation.

9.9.1   ROLE OF DESIGNERS IN IMPROVING RELIABILITY AND ACCURACY OF 3D PROTOTYPING

Designers can play an instrumental role in improving the reliability and accuracy of designs created using 3D prototyping software. First, most software developers rely on user feedback from designers to create effective software updates, and to support this they often host online forums and events to solicit information. It is therefore important for designers to support this level of collaboration to enable development and progress.

Second, the importance of maintaining practice with manual methods of design, cut, and make needs to be embraced so that it can inform the digital design process in terms of how things translate from digital to physical. This is particularly relevant when virtually designing features that are often hidden or overlooked during the digital design process such as internal linings, tapes, and interlinings. Third, employers need to support designers with ongoing professional development in the form of regular training and practice with 3D prototyping software which is essential as software updates can often contain new developments that enhance the reliability and accuracy of 3D prototyping.

9.9.2 Role of Software Developers in Improving Reliability and Accuracy of 3D Prototyping

The uptake of 3D prototyping within the apparel industry has been slow over the last two decades but has been accelerated by industry challenges, including the recent global pandemic, the drive towards sustainable design practices and ongoing developments in computer processing which have enabled 3D prototyping software to achieve enhanced photorealistic qualities (McKinsey, 2020). Feedback from user forums has been an important source of information supporting software improvements relating to design, cut, and make, but it is the software programmers working alongside experienced fashion designers and product developers who are the driving force of software improvements. Some of these focus on concurrently resolving small-scale glitches or bugs resulting in the release of a minor software update whilst others such as improving the simulation of human flesh for fit purposes are highly complex and longer-term projects often involving AI. Cumulatively, these developments are rapidly narrowing the gap between the virtual and the physical prototyping processes.

There are already recent examples of the experimental use of AI and machine learning for fashion design purposes in the luxury fashion sector such as Acne

FIGURE 9.8 ACNE Studio AI Fashion Collection (2021)

Source: Courtesy of www.acnestudios.com

Studio's AI Autumn 2020 collection, which was designed using a form of AI known as a Generative Adversarial Network (GAN) as opposed to a pattern-dependent 3D prototyping software.

It is anticipated that the gap between virtual and physical fashion design prototyping will be further narrowed with the support of AI to improve its accuracy but also to potentially automate 3D prototyping. According to Luce (2019), AI tools are becoming more accessible to a 'diverse set of people to train their own models with their own data without the help of machine learning researchers' (Luce, 2019, p. 185). This means it is now possible and affordable for software developers who don't have a background in AI or machine learning to more fully and more affordably integrate AI and machine learning to advance the accuracy and to even fully automate 3D prototyping software, thus inferring that the current 3D prototyping software platforms and market that we know today may be quite different in the near future.

REFERENCES

Ahmad, S, et al. (2017) *Advanced Textile Testing Techniques*, Routledge, CRC Press, Boka Raton, FL

Amerfird (2010) *Minimizing Seam Puckering* [online], www.amefird.com/wp-content/uploads/2010/01/Minimizing-Seam-Puckering-2-5-10.pdf (Accessed 02/01/2024)

BOF-McKinsey (2022) *The State of Fashion* [online], www.mckinsey.com/~/media/mckinsey/industries/retail/our%20insights/state%20of%20fashion/2022/the-state-of-fashion-2022.pdf (Accessed 02/01/2024)

BSI (1991) *BS 3870–1:1991, Stitches and Seams Classification and Terminology of Stitch Types* [online], www.en-standard.eu/bs-3870-1-1991-stitches-and-seams-classification-and-terminology-of-stitch-types/#:~:text=This%20International%20Standard%20classifies%2C%20designates,hand%20and%20machine%20sewn%20seams (Accessed 02/01/2024)

Coats (2022) *Seam Types* [online], www.coats.com/en/information-hub/seam-types (Accessed 02/01/2024)

Conn, R (2020) *How Artificial Intelligence Is Reshaping Generative 3D Modelling* [online], www.bricsys.com/en-eu/blog/how-artificial-intelligence-is-reshaping-generative-3d-modeling (Accessed 02/01/2024)

Guest, SF (2021) So What's Next for 3D in Fashion . . . And What Brands Must Demand from the Technology. *Sourcing Journal* [online], https://sourcingjournal.com/topics/technology/style3d-3d-zhejiang-design-studio-fashion-cloud-visualization-metaverse-317323/ (Accessed 22/12/2021)

Just-Style (2020) *Micro-Factories, the Future of Fashion Manufacturing* [online], www.just-style.com/features/micro-factories-the-future-of-fashion-manufacturing/?cf-view&cf-closed (Accessed 02/01/2024)

Luce, L (2019) *Artificial Intelligence for Fashion: How Artificial Intelligence Is Revolutionizing the Fashion Industry*, APRESS, San Francisco, CA

Mäkinen, H (2009) *Textiles for Cold Weather Apparel*, Woodhead, Cambridge

McKinsey (2018) *Skill Shift: Automation and the Future of the Workforce* [online], www.mckinsey.com/featured-insights/future-of-work/skill-shift-automation-and-the-future-of-the-workforce (Accessed 02/01/2024)

McKinsey (2020) *Fashion's Digital Transformation: Now or Never* [online], www.mckinsey.com/industries/retail/our-insights/fashions-digital-transformation-now-or-never (Accessed 02/01/2024)

McLoughlin, J and Mitchell, A (2015) Chapter 16 – Fabric Finishing, Joining Fabrics Using Stitched Seams. In: Sinclair, R (ed.) *Textiles and Fashion*, Woodhead Publishing, Cambridge

Postlethwaite, S (2022) *Reshoring UK Manufacturing with Automation, Royal College of Art & Innovate UK KTN/Made Smarter* [online], https://iuk.ktn-uk.org/wp-content/uploads/2022/03/Reshoring-UK-Garment-Manufacturing-with-Automation-Thought-Leadership-Paper-final-2.pdf (Accessed 02/01/2024)

Wang, QJ, et al. (2021) Getting Started with Virtual Reality for Sensory and Consumer Science: Current Practices and Future Perspectives. *Food Research International*, 145 [online], https://doi.org/10.1016/j.foodres.2021.110410 (Accessed 02/01/2024)

10 Advances in 3D Cloth Simulation and Prototyping for Apparel Design in Gaming

10.1 INTRODUCTION

This chapter will begin with a brief history of gaming, an outline of apparel creation in gaming before 3D prototyping software and cloth simulation, followed by a timeline outlining major advances that have supported the development of apparel prototyping and cloth simulation for the gaming industry. The chapter will move on to consider the impact of 3D prototyping software on apparel design for gaming. The chapter will finally consider the emergence of key fashion and gaming collaborations and the benefits for both sectors currently and in the near future.

To provide the most current insight regarding recent and emerging advancements in this area, the key contributors include Adam Cain (Lecturer of Games Art, School of Digital Arts at Manchester Metropolitan University), as well as a number of other contributors including Harvey Parker (Art Director at the Multiplayer Guys) and Shiloh Ragins (Lecturer of Games Art at Confetti Institute of Creative Technologies at Nottingham Trent University).

10.2 POTTED HISTORY OF GAMING

Video games have come a long way since the release of games like Pong in 1972. From the era of 8-bit and 16-bit pixels and the rise of arcade machines in the 1980s to the use of 3D modelling techniques and powerful graphics cards in the 1990s. The video game industry is constantly evolving and innovating with new technological advancements. During the turn of the millennium video games started to become what we see them as today, a fully integrated entertainment system on a home console or PC, with global social interconnectivity online and immersive 3D worlds utilising cutting-edge 3D graphics and techniques with constant improvements to the visual representation of real-world physics, textures, and lighting.

The games industry is worth an estimated US$300 billion, with 2.7 billion active players, and is growing at a rapid pace (Accenture, 2021). The growth of the gaming industry has been phenomenal within the last few years, part of this growth can be attributed to the global lockdown measures of COVID-19 in 2020, which led to a 30% increase in people playing video games more than five hours a week (Simon-Kucher & Partners; Dynata, 2020), and a 2022 study found that 71% of respondents,

DOI: 10.1201/9781003126454-10

over the age of 16, had increased the duration they play games during the COVID-19 outbreak and lockdown measures (Barr and Copeland-Stewart, 2021).

The fashion industry has capitalised on this growing market and further assisted to bring gaming culture into the mainstream with several key fashion brands developing partnerships with major gaming enterprises such as the Louis Vuitton and *League of Legends* (Riot Games, 2006) collaboration in 2019 (Louis Vuitton, 2019). One of the most popular gaming franchises *Call of Duty* (Activision, 2022) has generated US$30 billion in its lifetime revenue (Hume, 2022) and has also developed a wide range of branded merchandise and apparel. High-street retailers such as House of Fraser and Sports Direct have licensed *Call of Duty* imagery for a range of apparel whilst another popular independent fashion label Valaclava, a cyber-physical NFT game wear brand, has worked in collaboration to produce limited-edition apparel for the *Call of Duty* franchise to sit along with their Infinity Collection (Valaclava, 2022).

FIGURE 10.1   Valaclava Infinity Collection (2022); VLCV Enchantress Dress.

Source: Courtesy of www.Valaclava.com

FIGURE 10.2 Valaclava Infinity Collection (2022); VLCV Hunter Hoodie and Palladin Pants.

Source: Courtesy of www.Valaclava.com

The popularity of gaming has also led to the rise of Esports, which are competitive game-based tournaments that are streamed online and are now estimated to be worth approximately US$1.8 billion per annum (Accenture, 2021). These events have attracted a large audience and generated valuable sponsorship and investment deals with luxury fashion brands such as Gucci's 2020 collaboration with the global Esports organisation Fnatic (Fnatic, 2020).

Video game developers are increasingly using forms of monetisation for in-game products in their games, and it has overtaken video game purchases as the main source of income for developers (Statista, 2022). With video game players spending more time and money on in-game purchases, the representation of their personal style within digital spaces via customised avatars, skins, and digital clothing has become both lucrative and popular for developers and players alike (Sullivan and Caldwell, 2021). As this trend continues to grow, so will the demand for popular fashion retailers and designers to provide digital clothing for people online, as can already be seen with the rise in collaborations between high-end fashion retailers and popular gaming franchises, such as the 2021 collaboration with *Gucci* and

Roblox (Roblox Corporation, 2006) or the 2022 *Balmain* and *Pokemon* (Nintendo, 1996) collaboration.

During the recent COVID-19 pandemic, fashion designers have been able to tap into new markets and create products that appeal to a wider range of customers previously unexplored. A recent and notable collaboration was the 2021 partnership with luxury fashion brand Balenciaga and the global gaming phenomenon *Fortnite* (Epic Games, 2017). The physical Balenciaga hoodie with *Fortnite* branding was sold out at US$725 whilst the digital Item Shop Bundle of Balenciaga-designed skins sold on the Epic store for 1,800 V-bucks (a form of digital currency used in *Fortnite* worth approximately US$14 for the bundle). The *Fortnite* YouTube account currently has 11.3 million subscribers, and the advertisement for this collaboration has been viewed more than 500,000 times (Fortnite, 2021). Epic also claims that *Fortnite* has over 350 million registered users (Fortnite Game, 2020), evidencing how powerful this form of promotion and advertising can be in developing brand loyalty for fashion and apparel production companies.

Whilst most of the fashion and gaming partnerships are in the form of collaborative promotion for real-world purchases, there is now a growing social shift where digital lives and real lives coexist, whether that is in the form of social media or online playing video games. Given these current trends and growth in the games and fashion collaboration market, it is anticipated that the fashion industry will continue to pursue digital worlds as a platform for both advertising of real-world clothing and digital apparel consumerism. Lyst's Digital Fashion Report in 2021 provides evidence for the impact of video games on marketing and promotion showing a 41% increase in searches for Balenciaga on Lyst, 48 hours after the release of *Balenciaga—Afterworld: The Age of Tomorrow* video game (Lyst, 2021). Fashion designers are no longer designing garments just for our physical appearance but also contributing to how people and their digital avatars look in different gaming worlds. As such, better cloth simulation, which is the process of calculating all aspects of what clothes and fabrics do in real life, including aspects like movement, stretching, folding, and wrinkles and applying it to a 3D environment or 3D model, is essential to create a greater sense of authenticity and realism to the illusion of physics, textures, and lighting in these digital worlds.

10.3 APPAREL CREATION IN GAMING BEFORE THE INTRODUCTION OF 3D PROTOTYPING SOFTWARE AND CLOTH SIMULATION

The 1990s was the true dawn of 3D video games, as the technological advancements in hardware such as CD-ROMS and Graphical Processing Units (GPUs) found in consoles such as *Sony Computer Entertainments Sony PlayStation* in 1994 and the *Sega Dreamcast* in 1998 allowed for the industry to shift from 2D graphics to 3D outputs and support the creation of more realistic 3D characters and clothing with polygons, texturing, and lighting. This switch to 3D graphics meant that video games could begin to take on a more realistic appearance, and it allowed game developers to utilise a range of techniques to integrate pre-rendered and real-time 3D graphics into games.

The representation of clothing and apparel within 3D video games of the 1990s was rudimentary by today's graphical standards. Texturing in video games became extremely important in heightening the realism and readability of specific materials in games. When creating elements for games such as a character in costume or apparel, the geometry had to be split into separate components and attached with skeleton rigs. Artists would spend a great deal of time giving the illusion of folds in fabric, textured surfaces, and lighting. This is particularly notable in highly successful games such as *Resident Evil 2* (Capcom, 1998).

Textures for apparel in video games were created using texture maps, which can be done by scanning real-world textures or by manually creating them in a 2D digital program. Once the texture map is complete, it can be used to wrap around the object or environment to give it a more dynamic and enhanced appearance (Heckbert, 1986). As 3D modelling techniques improved, so did the application of textures and lighting to character clothes and apparel, with techniques such as bump maps, normal maps, displacement maps, and shaders. As games moved into the second generation of 3D-capable consoles and hardware in the late 1990s and early 2000s, there was a notable improvement in the resolution of texture and lighting (Benno, 2019), both allowing for a greater fidelity of replicating apparel in games. In 3D modelling, bump mapping is used to create an illusion of texture detail and depth. Textures are artificially added to a 3D model using greyscale 2D imagery and lighting rather than applying it directly to the 3D model (Pluralsight, 2022), thus saving space for the optimisation of a video game. Further development in this form of texturing is the use of normal mapping, which can be considered a significant advancement in the creation of high-quality 3D graphics for video games. It allows game developers to use low polygon models that respond to the lighting with more detail than is in the actual model and create the illusion of high polygon, more detailed models (Pluralsight, 2022).

Most games will now stipulate a minimum system requirement to customers before they purchase a game which will detail the specific graphics card or platform required to play. This limitation means that a high polygon character design created in software such as *Z-Brush* (Pixologic, 2003), a 3D sculpting software, and *Marvellous Designer* (CLO3D, 2009), a digital cloth making program will need to be baked onto a low polygon model. The process of baking allows for the texture and lighting detail to be transferred over to less processor-intensive models. Many materials are now created and applied using software such as *Substance Designer* (Adobe, 2019). *Substance Designer* has been another revolutionary piece of software for the games industry and any industry using 3D models as it allows a creator to use procedural, pre-made, hyper-realistic textures. When working in the fashion industry, it enables artists to save time by exploring the look of materials for clothes without having to pay the associated costs of finding materials and manufacturing (Shaw, 2020). Whilst many of these techniques and practices are still used today in the production of texture, apparel, and cloth in video games, the industry is moving towards programmable real-time cloth simulation to calculate and imitate real-world physics of materials in real time within a gaming world, as seen in Epic's 2022 *Unreal Engine 5* release (Epic Games, 2022).

TABLE 10.1

Timeline of Key Developments in Gaming Relating to 3D, Virtual Clothing, and Cloth Simulations

1988	SoftImage3D (Softimage, 1988)
1994	Sony PlayStation (Sony Computer Entertainment, 1994)
1996	Nintendo 64 ((Nintendo Co,. Ltd, 1996)
1996	Direct 3D (Microsoft Corporation, 1996)
1996	Side FX Houdini FX (Side Effects Software Inc, 1996)
1996	3dfX Voodoo (3dFx Interactive, Inc, 1996)
1996	Autodesk 3DS Max (Audodesk, Inc, 1996)
1998	Maya (Alias Systems Corporation, 1998)
1999	Nvidia Geforce 256 (Nvidia Corporation, 1999)
1999	Z-Brush (Pixologic, 1999)
2000	SoftImage XSI (Softimage, Co, 2000)
2001	Havok SDK (Havok, 2001)
2002	SyFlex—Cloth Simulator (SyFlex, 2002)
2004	Valve Source Physics Engine (Valve Corporation, 2004)
2005	Xbox 360 (Microsoft Corporation, 2005)
2006	PlayStation 3 (Sony Computer Entertainment, 2006)
2008	Havok Cloth (Intel Corporation, 2007)
2009	Unreal Engine 3 (Epic Games, Inc, 2009)
2009	Cry Engine 3 (Crytek GmbH, 2009)
2009	Marvellous Designer (CLO3D, 2009)
2011	NIVIDA Apex Clothing (Nvidia Corporation, 2011)
2013	PlayStation 4 (Sony Computer Entertainment, 2013)
2013	Xbox One (Microsoft Corporation, 2013)
2014	Unreal Engine 4 (Epic Games, Inc, 2014)
2015	Ubisoft Motion Cloth (Ubisoft, 1986)
2017	Nintendo Switch (Nintendo Co,. Ltd, 2017)
2017	Cry Engine 5 (Crytek GmbH, 2017)
2018	Side FX Houdini 17—Vellum (Side Effects Software Inc, 2018)
2020	PlayStation 5 (Sony Computer Entertainment, 2020)
2020	Xbox Series X (Microsoft Corporation, 2020)
2021	SWISH Neural Network Cloth Simulation (Electronic Arts Inc., 2021)
2022	Unreal Engine 5 (Epic Games, Inc, 2022)

10.4   IMPACT OF 3D PROTOTYPING SOFTWARE ON APPAREL DESIGN FOR GAMING

In 2009, *CLO3D* launched its first beta release of *Marvellous Designer*, a dynamic 3D software program that allows for realistic cloth simulation. Originally intended as a solution for the visualisation of apparel in the entertainment industry, Marvellous Designer quickly became the industry standard tool for video game developers hoping to create authentic-looking apparel or cloth simulation in a 3D environment (Cartoon Brew Connect, 2021). This also applies to other textures and surfaces applied to clothes such as weathering, dirt, moisture, or damage. Cloth simulation is more than creating a single image of realistic apparel in 3D software, it is the animation of realistic authentic-looking apparel (80 Level, 2020). Combined with other cutting-edge technology such as performance capture and high-definition 3D graphics, video games such as *Cyberpunk 2077* (CD Projekt Red, 2020) began to take on genuine cinematic qualities. As it is now possible to create an incredibly detailed character on a single mesh, an artist can create clothing directly in *Marvellous Designer* (CLO3D, 2009) to generate virtual apparel.

One of the major selling points of cloth simulation is that it allows the user to create reusable custom assets changing the style and materials of a garment quickly to suit the vision of the intended output whether that be for video games or for real-world garments. This is particularly important for the modern-day fashion and apparel industries struggling with the impact of environmental sustainability and waste reduction during the current climate crisis (Edited, 2020). Cloth simulation is particularly important for modular design; when designing outfits within games, artists need to consider how players can pair up different items and customise their characters. Apparel and accessories must all work well together aesthetically but also technically within the game engine and avoid collision errors or clipping, which is when 3D objects collide and obscure the view of the object. Many fashion designers are now using the 3D pipeline for the creation of outfits and designs. In 2010, *CLO3D* released the first version of *CLO*, a 3D fashion design software like *Marvellous Designer* (CLO3D, 2009) but with a specific focus on the apparel industry. *CLO* gives the user the ability to create virtual prototypes and patterns that can support on-demand manufacturing and aspects of sustainable design practice.

Whilst *Marvellous Designer* is an effective tool for cloth simulation, there is a whole range of issues with using the actual cloth simulation mesh in terms of optimisation for games. Often, the designs will be imported into other software for retopology, which is the process of cleaning up and simplifying a mesh's topology for use in other applications, as the original output is often too expensive to run within games at 60 frames per second (FPS) (Cartoon Brew Connect, 2021). There are other issues, specifically centred around animation and optimisation at 60FPS, as each frame of animation using *Marvellous Designer* needs to be exported out as singular files. This is extremely labour-intensive on the GPU and processing capabilities of the console a game is running on. This means video game developers need to use various tricks and techniques to overcome this barrier such as applying cloth simulation to only important assets. Whilst there have been many important technological developments in cloth simulation such as *SyFlex*'s (Syflex, 2002) cloth simulator released in 2002, *Havok Cloth* (Havok, 2008) in 2008, and *NVIDIA Apex* (Nvidia Corporation, 2008) released in 2011. It is still relatively new and is continuing to evolve; as technology improves, so will the possibilities of what can be made in both digital and physical apparel.

10.5 MERGING OF APPAREL DESIGN FOR GAMING AND APPAREL DESIGN FOR FASHION

Major developments in cloth simulation technology have presented new opportunities to create virtual clothing for the games industry that also has relevance to high-street consumers. Marketing teams have developed important licensing opportunities on both sides of the industry such as *Square Enix*'s 2016 collaboration with *Roen* for the apparel design of lead characters in *Final Fantasy XV* (Square Portal, 2016). In 2012, Prada envisioned what characters from *Final Fantasy XIII-2* (Square Enix, 2012) would look like in its spring and summer collection.

During the pre-production phase of game development, artists and designers will work to visualise ideas for specific assets. In the case of a character, a concept artist will be tasked with designing not only the physical form of the character but also the apparel it wears and accessories it uses in various design styles. Artists working within this stage take design inspiration from all aspects of life, and depending on the overall game theme, mood, and style, an artist will often take inspiration from recent, current, or predicted fashion trends (Mulrooney, 2012). It is important that the designs of apparel in games with customisable avatars or outfits are what the audience will want to use for their avatar because the more popular a particular form of apparel or *skin* is within the game, the more likely it can be sold as a digital asset and the higher potential for it to become a physically manufactured form of merchandise that a consumer can wear in real life as discussed in Statista's 2022 report 'Gaming Monetization Statista Dossier'.

CyberPunk 2077 (CD Projekt RED, 2020), a science-fiction game, is one example of a game with many influences from contemporary apparel production. *Marvellous Designer* (CLO3D, 2009) was used extensively in *CyberPunk 2077* (CD Projekt RED, 2020) for the production of apparel and accessories (Cartoon Brew Connect, 2021). Graphics and logos found within *CyberPunk 2077* are now available directly from the CD Projekt RED website as T-shirts, exemplifying the consumers' desire to share their passion for a favoured game in the real world (CD Projekt Red Gear, 2022). As the video games industry has developed, licensing deals for specific video games have been granted to small companies specialising in video game-branded merchandise, often for niche customer bases. Licensing has now evolved into a substantial additional income for video game developers and a secondary source of promotion (Gamex Studio, 2022).

Traditional high-street fashion brands are now regularly seeking video game licences or collaborations, as had been done previously for famous bands and movies. For example, in 2016 Moschino collaborated with *Nintendo* (Nintendo Life, 2015) to celebrate the 25-year anniversary of the *Super Mario Bros* franchise. This was one of the first collaborations between a luxury fashion brand and a video game developer which established a trend now firmly set in the cultural zeitgeist, one that shows video games are no longer a niche market reserved for specific subcultures but a global digital community of people who enjoy the worlds of both fashion and video games. *Roblox* (Roblox Corporation, 2006), an online gaming platform for creating your own games and worlds, is now a leading collaborator with fashion and apparel brands, with over US$1 billion spent on virtual goods, and 54% of the average daily users being 12 or under it is obvious that this is not only a lucrative market but also an influential platform for building customer loyalty (Influencer Marketing Hub, 2022). Recent collaborations in 2021 with *Roblox* have also included *Gucci Garden*, a virtual environment hosting a range of purchasable digital content co-created by Gucci. Since 2021, virtual game environments hosting fashion and apparel events have become a normal occurrence; in 2021, Balenciaga released its first game *Afterworld: Age of Tomorrow*, a multiplatform adventure game and marketing showpiece for its Fall 2021 collection. This game is a precursor to many other future titles and collaborations between game developers and apparel manufacturers embracing what is now commonly referred to as the metaverse, evidencing that famous high-end fashion apparel is no longer solely for the physical world (Future of Marketing Institute, 2022).

10.6   THE FUTURE AND CONCLUSIONS

The future relationships between the gaming industry and the fashion industry are likely to be further developed by new digital approaches to the consumption of apparel such as Non-Fungible Tokens (NFT) and ownership. In 2021, Dolce and Gabbana released its first NFT collection which sold for US$6 million (Thomas, 2021). Customers were able to purchase both physical and digital garments for use on digital gaming platforms and social media. The metaverse and Web 3.0 is predicted to take these developments further with companies continuing to embrace the power of digital youth culture in the form of video games and social media. Most notably, games like *Fortnite* (Epic Games, 2017), *Minecraft* (Mojang 2011), and *Roblox* (Roblox Corporation, 2006) are generating huge revenues from in-game ownable assets. One of the main issues with this form of *ownership* is that you can use the digital asset only on the specific platform it was purchased in. As NFTs become more readily available across various platforms in the form of blockchain, and games and digital cryptocurrency become adopted more broadly within society, it is hypothesised that a consumer will be able to purchase digital apparel from a specific company and that ownership will transcend singular platforms and be readily available across all digital space.

In April 2022, *Epic Games* released *Unreal Engine 5* for all game creators, this game engine is groundbreaking in its capacity for representing a new level of realism for cloth simulation in a digital environment. *Epic Games* has integrated a cutting-edge system called *Nanite*, which supports higher tri/poly count, with special inbuilt support for cloth simulation (Unreal Engine, 2021). This will enable people to experience cloth simulation bordering on true realism when playing games or experiencing virtual environments, whether that be in virtual reality, augmented reality, film, videos, live performance, or traditional forms of gaming on PC or console. Physical real-world photography or advertising can now be replaced by completely digitally rendered marketing campaigns indistinguishable from real garments or apparel as discussed and evidenced in *Vogue* Taiwan's 'Future in Transit' (Vogue, 2020).

Currently, there are many companies developing highly sophisticated algorithms and exploring the potential of machine learning and AGI (artificial general intelligence) in game development and cloth simulation. AGI is now capable of developing highly detailed and realistic cloth simulations for games, such as EA Seeds SWISH—Neural network cloth simulation in Madden NFL 21 (Electronic Arts, 2021). As these programs develop and improve, game developers or manufacturers of apparel using 3D prototyping will no longer need to solve the problems of current cloth simulation tools but look to improve the algorithms of the AGI cloth simulation programs. Machine learning and AGI are already being used in the fashion industry for identifying trends and customer user experience, body mapping, styling, waste reduction, etc. such as *Stylumia's Consumer Intelligence Tool* (Stylumia, 2022), but it is also being used for design and the simulation of high-end designs, reducing costs of iteration and runs for physical products as seen with the 2016 *Zalando* and *Google* collaboration *Project Muze* (Google, 2016). Given the improvements in machine learning and AGI creativity found in programs such as *Open AIs' DAll-E 2* (Openai, inc, 2020), it is not hard to imagine a future whereby AGI can be tasked with all aspects of apparel manufacturing and production for both the physical world and the digital sector, from initial concept, design, prototyping, simulation, manufacturing, distribution, and promotion (The Tech Fashionista, 2021).

The fashion and video game industries have never been more interconnected than they are now. As game developers continue to pioneer revolutionary ways to simulate hyper-realistic detailed cloth simulation and utilise their platforms for the marketing and monetisation of digital and real-world assets, the fashion industry will continue to benefit from these collaborations and the technology available. The continued evolution of 3D graphics, 3D prototyping, Web 3.0, and machine learning will push the fashion industry into a new era that transcends the physical world and allows customers to celebrate their love and loyalty to specific brands and designs in real life and online simultaneously. These new technologies will provide revolutionary creative freedom and allow people to design and create with more autonomy, exuberance, and innovation than ever before.

REFERENCES

Accenture (2021) *Gaming: The Next Super Platform* [online], www.accenture.com/us-en/insights/software-platforms/gaming-the-next-super-platform#:~:text=2.7%20billion%3A%20the%20estimated%20number,higher%20than%20previous%20industry%20estimates (Accessed 02/01/2024)

Barr, M and Copeland-Stewart, A (2021) Playing Video Games During the COVID-19 Pandemic and Effects on Players' Well-Being. *Games and Culture*, 17(1) [online], https://journals.sagepub.com/doi/full/10.1177/15554120211017036 (Accessed 02/01/2024)

Benno R (2019) *Twitter*, 29 April 2019 [online], https://twitter.com/BryanRenno/status/1122880481643487232 (Accessed 02/01/2024)

Cartoon Brew Connect (2021) *How Marvelous Designer Helped Bring the Immersive Dystopia of Cyberpunk 2077 to Life* [online], www.cartoonbrew.com/sponsored-by-marvelous-designer/how-marvelous-designer-helped-bring-the-immersive-dystopia-of-cyberpunk-2077-to-life-210422.html (Accessed 02/01/2024)

CD Projekt Red Gear (2022) *CD Projekt Red Gear* [online], https://gear.cdprojektred.com/ (Accessed 02/01/2024)

Edited (2020) *Digital Meets Physical: Defining the Retail Landscape in 2020* [online], https://edited.com/blog/digital-meets-physical-defining-the-retail-landscape-in-2020/ (Accessed 02/01/2023)

Electronic Arts (2021) *Siggraph 21: Swish—Neural Network Cloth Simulation in Madden NFL 21* [online], www.ea.com/seed/news/swish-neural-network-cloth-sim-in-madden-nfl-21 (Accessed 02/01/2024)

Epic Games (2022) *Cloth Simulation in Unreal Engine* [online], https://docs.unrealengine.com/5.0/en-US/cloth-simulation-in-unreal-engine/ (Accessed 02/01/2024)

Fnatic (2020) *Fnactic X Gucci* [online], https://fnatic.com/news/fnatic-x-gucci (Accessed 02/01/2024)

Fortnite (2021) *Balenciaga Brings Digital Fashion to Fortnite* [online], www.epicgames.com/site/en-US/news/balenciaga-brings-high-fashion-to-fortnite (Accessed 02/01/2024)

Fortnite Game (2020) *Twitter*, 6 May [online], https://twitter.com/FortniteGame/status/1258079550321446912 (Accessed 02/01/2024)

Future of Marketing Institute (2022) *The Future of Digital Fashion Is Digital* [online], https://futureofmarketinginstitute.com/the-future-of-fashion-is-digital/?utm_source=rss&utm_medium=rss&utm_campaign=the-future-of-fashion-is-digital (Accessed 01/12/2022)

Gamex Studio (2022) *Video Game Merchandise Industry from AAA Studios to Indies* [online], https://gamex-studio.com/video-game-merchandise-industry-from-aaa-studios-to-indies/ (Accessed 02/01/2024)

Google (2016) *Project Muze: Fashion Inspired by You, Designed by Code* [online], https://blog.google/around-the-globe/google-europe/project-muze-fashion-inspired-by-you/#:~:text=Project%20Muze%2C%20an%20experiment%20from,the%20inspiration%20for%20unique%20designs (Accessed 02/01/2024)

Heckbert, P (1986) Survey of Texture Mapping. *IEEE Computer Graphics and Applications*, 6(11) [online], https://ieeexplore.ieee.org/document/4056764 (Accessed 27/12/2023)

Hume, M (2022) The Future of Call of Duty and 'Warzone'. *The Washington Post*, 8 June [online], www.washingtonpost.com/video-games/2022/06/08/call-duty-future-modern-warfare-2-warzone-2/ (Accessed 02/01/2024)

Influencer Marketing Hub (2022) *55 Amazing Roblox Statistics Revenue, Usage & Growth Stats* [online], https://influencermarketinghub.com/roblox-stats/ (Accessed 02/01/2024)

Louis Vuitton (2019) *Louis Vuitton X League of Legends* [online], www.vogue.com/article/louis-vuittons-new-capsule-with-league-of-legends (Accessed 02/01/2024)

Lyst (2021) *The Digital Fashion Report—Lyst X The Fabricant* [online], www.lyst.com/data/digital-fashion-report/ (Accessed 02/01/2024)

Mulrooney, M (2012) *Interview in Conversation with Claire Hummel Concept Artist Bioshock Infinite* [online], https://alternativemagazineonline.co.uk/2013/06/06/interview-in-conversation-with-claire-hummel-concept-artist-bioshock-infinite/ (Accessed 02/01/2024)

Nintendo Life (2015) *Nintendo Smartens Up For "Super Moschino" Collection of Shirts, Sweaters and Accessories* [online], www.nintendolife.com/news/2015/12/nintendo_smartens_up_for_super_moschino_collection_of_shirts_sweaters_and_accessories (Accessed 02/01/2024)

Pluralsight (2022) *Difference Between Displacement, Bump and Normal Maps* [online], www.pluralsight.com/blog/film-games/bump-normal-and-displacement-maps#:~:text=As%20we%20already%20know%2C%20a,Z%20axis%20in%203D%20space (Accessed 02/01/2024)

Roblox Corporation (2006). *Roblox*. Roblox Corporation [online], www.roblox.com/ (Accessed 10/11/2022)

Shaw S (2020) *Substance by Adobe Helps Keep INDG's Product Visualization Business in the Pink* [online], https://blog.adobe.com/en/publish/2020/07/16/substance-by-adobe-helps-keep-indgs-product-visualization-business-in-the-pink (Accessed 02/01/2024)

Simon-Kucher & Partners; Dynata (2020) *New Study: Gamers Around the World Are Spending More Time and Money on Video Games during the COVID-19 Crisis, and the Trend Will Likely Continue Post-Pandemic* [online], www.simon-kucher.com/en/insights/global-gaming-study-more-gamers-spending-more-money-covid-lockdowns-which-publishers-will (Accessed 02/01/2024)

Square Enix (2012) *Final Fantasy Characters Showcase Prada 2012 Men's Spring Summer Collection* [online], www.square-enix-games.com/en_GB/news/final-fantasy-characters-showcase-prada-2012-mens-springsummer-collection (Accessed 02/01/2024)

Square Portal (2016) *A Closer Look at Roen x Final Fantasy XV Fashion* [online], https://squareportal.net/2016/04/01/closer-look-at-roen-x-final-fantasy-xv-fashion/ (Accessed 02/01/2024)

Statista (2022) *Gaming Monetization Statista Dossier* [online], www.statista.com/study/41424/gaming-monetization-statista-dossier/ (Accessed 02/01/2024)

Stylumia (2022) *Consumer Intelligence Tool* [online], www.stylumia.ai/our-solutions/consumer-intelligence-tool/ (Accessed 02/01/2024)

Sullivan, M and Caldwell, A (2021) *The Spending Habits of MMO Gamers*, *Top Dollar* [online], www.accredited debtrelief.com/blog/mmo-money-mmo-problems/ (Accessed 10/11/2022)

The Tech Fashionista (2021) *AI in the Fashion Industry: 6 Game-Changing Technologies* [online], https://thetechfashionista.com/ai-in-the-fashion-industry/ (Accessed 02/01/2024)

Thomas, D (2021) *Dolce & Gabbana Just Set a US$6 Million Record for Fashion NFTs* [online], www.nytimes.com/2021/10/04/style/dolce-gabbana-nft.html (Accessed 02/01/2024)

Unreal Engine (2021) *Nanite in UE5: The End of Polycounts*, Unreal Engine [online], www.youtube.com/watch?v=xUUSsXswyZM (Accessed 02/01/2024)

Valaclava (2022) *Our Story, Valaclava, Cyber Physical NFT Apparel* [online], https://valaclava.com/pages/story (Accessed 02/01/2024)

Vogue Taiwan (2020) *Vogue's First Computer-Generated Virtual Cover—Future in Transit* [online], www.vogue.com.tw/fashion/article/2020-may-cover (Accessed 02/01/2024)

11 Advances in Product Lifecycle Management (PLM) Systems within the Fashion Industry

11.1 INTRODUCTION

This chapter will open by providing an overview of the development of Product Lifecycle Management (PLM) including its implementation within the fashion industry, summarised as a timeline of key advances. The chapter will move on to consider leading vendors of fashion PLM systems and the relevance and value of this technology to the post-pandemic fashion industry by contextualising recent developments and advances that are driving this technology including post-pandemic working practices/post-pandemic consumer, sustainability and transparency for social compliance (using scientific impact calculations), Bill of Labour, interoperability: design-focused integrations and plug-in technologies and licensing solutions that support collaboration with partners in the value chain. The chapter will close with a consideration of future PLM advances. To provide the most current insight into advances in PLM software the key contributor for this chapter is Mark Harrop, often dubbed the PLM guru, who is a highly respected fashion industry PLM expert and CEO of a leading industry-focused PLM market analysis publication WhichPLM. Other contributors include Gillian Pinkhardt, Global Marketing Director Coats Digital.

11.2 THE ORIGINS OF PLM

According to Luciano (2010), the first PLM software system was developed by the American Motors Corporation (AMC) during the 1980s, when the company was struggling to compete with larger competitors. In an attempt to compete more efficiently within the wider car manufacturing industry, AMC decided to invest in a system that could track a product across its lifecycle to identify opportunities to reduce waste and improve efficiencies. Their system proved to be effective and enabled the company to grow its market share to the point where it was bought by Chrysler and became the car industry's lowest cost-producer by the mid-1990s (SAP, 2022).

In general terms, PLM is a system or software platform that is used to improve the way that products are managed from concept to delivery by integrating all the product data, processes, business systems, materials, and, ultimately, people who will be operating across the extended value chain. PLM connected with Enterprise

Resource Planning (ERP) is used for the management of business processes such as live tracking of production lines in manufacturing and other best-of-breed business systems making it possible to follow the lifecycle of each component and product, from concept to consumer and finally to a product's disposal/recycle.

PLM can therefore be used for smarter decision-making, based upon real-time data feeds, highlighting problems, improving efficiencies, delivering faster product development, increasing sales at higher margins, and reducing the overall costs of goods sold (COGS).

11.3 ORIGINS OF PLM IN THE FASHION INDUSTRY

PLM started its life in the fashion industry in late 1999 by Freeborders Inc. in San Francisco which designed and developed a system called Collaborative Product Management (CPM), which was later changed to PLM, which it marketed in 2002 (Harrop, 2022a). PLM was designed to provide a single organised version of the 'facts' for the foundational data of a product and became increasingly important in the 1990s when globalisation, in search of lower labour cost, forced more and more fashion businesses to offshore manufacturing to remain competitive, and this directly necessitated the need for advanced international communications (Segonds et al., 2014; McCormick et al., 2014).

The early adopters of what was then known as Product Data Management (PDM) started to communicate with each other using product specifications, known today as tech packs. This limited level of data was adequate for most fashion businesses at this time because few had more than four collections per year with a limited number of suppliers, and the majority retailed via bricks and mortar stores (PLM Report, 2022a). The introduction of PDM enabled the sharing of data initially by fax and eventually moved onto web-enabled technologies like Citrix and PC-Anywhere. These were the most reliable tools to communicate across the internet at this point in time, as email and the internet were not widely available until the mid-1990s. From the 2000s, the practices of PDM had recognised limitations, unable to support the complex workflow management processes, and collaboration demanded from fashion businesses who were now operating in a fast fashion business environment involving offshore manufacturing and retailing via e-commerce.

Table 11.4 provides a summary of some key developments of PLM for the fashion industry.

11.4 PLM FOR TODAY'S FASHION INDUSTRY

PLM systems have been adopted by many of today's fashion businesses to manage products through their lifecycle, supporting improvements in speed to market, quality, increased efficiencies, and cost benefits (Conlon, 2020; Harrop, 2022a).

Prior to the millennium period, the fashion product lifecycle was simpler to manage than today as it involved a linear journey from concept to disposal that could be managed using basic tools such as Excel spreadsheets and email. However, in the last ten years this journey has become much more complicated. The post-pandemic fashion industry is adapting to the evolving values of today's consumers which is driving the

TABLE 11.1
Timeline of Key PLM Advances in the Fashion Industry

1987	Microdynamics (Dallas Texas) developed the world's first fashion PDM solution.
Mid-1990s	PDM was reaching its limits of an on-premise client–server application.
1994	Gerber Technology acquired Microdynamics and with this acquisition came PDM, which was already being marketed as WebPDM, this was to be the last version of PDM developed by Microdynamics.
1999	Harrop designed a blueprint for the next-generation CPM and presented it to the Gerber board, unfortunately the vote went against building CPM.
1999–2003	Freeborders developed CPM which was renamed PLM and released it in 2003.
2005	Over 500 retailers, brands, and textile businesses use the first generation of PLM.
2005	In 2005, PTC acquired Aptavis and went on to configure the PLM solution for other market segments such as apparel, fashion, luxury, retail, sports/performance wear, and so on. That PLM solution was rebranded as FlexPLM and is now the outright global market leader in these sectors.
2003	Aptavis Technologies (a PTC partner) developed the first web-based PLM solution for the footwear industry. The first customers were Nike, Reebok, and Timberland.
2012	Cloud-based PLM as a SaaS model was introduced by several fashion PLM vendors.
2012	PTC introduced the first bidirectional integration between FlexPLM and Adobe Illustrator.
2013	PTC was the first PLM company to enable brands and retailers to aggregate data from PLM and other business systems (IoT).
2015	PTC was the first PLM company to enable brands and retailers to apply AI/machine learning to its product development data to drive better business decisions using predictive analytics and create a *Closed Loop* PLM approach.
2018	3D sample review mobile app launched by Centric PLM.
2018	Coats Digital VisionPLM launched the first bidirectional PLM extension on Adobe Exchange.
2019	Centric PLM pioneers a fully digital, 3D design and development workflow with integration to multiple leading 3D fashion solutions such as CLO, Browzwear, and EFI Optitex
2020	Coats Digital Launched GSDCost with Fair Wage Tool supporting PLM BOL.
2020	Clo-Vise Plugin was developed by PTC Flex PLM and CLO3D partnership.
2021	PTC was the first PLM provider to enable users to 'optimize' 3D assets directly in PLM that enabled sharing internally and across the extended supply chain. It was also the first vendor to offer 3D asset visualisation, review, and markup capabilities directly in FlexPLM (Zaczkiewiez, 2021).
2022	3D Smart Connector developed by Bamboo Rose PLM.
2022	Adobe Illustrator Plugin implemented by Backbone PLM.
2022	PLM considered an essential tool for managing new directives around human rights and environmental due diligence.
2022	Made2Flow, a company that specialises in scientific measurement of environmental indicator's partnered with PLM leader called Inform-PLM.
2022	PTC was the first PLM company to integrate with MakerSights technology, enabling brands to bring consumer feedback directly into the product development process.

Note: This timeline relates to only retail, footwear, and apparel industry.

launch of new style options, services, and varieties at a much faster pace. The emphasis of the consumer has led to the development of new start-ups, such as MakerSights, that have enabled the automation of consumer data sharing via PLM into key areas of the value chain such as product development, merchandising, and go-to-market workflows.

In addition to meeting more complex consumer requirements, there is also the added pressure of producing fashion products in a leaner and more cost-effective way whilst delivering transparency and sustainability across the value chain. The fashion product lifecycle has therefore evolved from a linear to a circular one, and the management of this more complicated lifecycle can't be achieved without the use of an integrated, multifunctional, enterprise-wide PLM solution (Harrop, 2022a; The Interline, 2021; Conlon, 2020). The purpose of today's PLM systems is eloquently captured by Corallo:

> [A] strategic business approach that supports all the phases of product lifecycle, from concept to disposal, providing a unique and timed product data source. Integrating people, processes and technologies and assuring information consistency, traceability and long term archiving, PLM enables organisations to collaborate within and across the extended enterprise.
>
> (Corallo et al., 2013, p. 6)

PLM in today's increasingly digitalised fashion industry serves as a central hub that enables real-time collaboration and decision-making along with the sharing of accurate, actionable data, and assets whenever and wherever they're needed, reshaping the industry to be 'more sustainable, more intelligent, more efficient, and more transparent' (McCready-Stocks, 2022, p. 62). PLM is supporting today's fashion businesses to develop sustainable and socially responsible products faster, at the right price and quality whilst eliminating waste, loss of revenue and reputation associated with incomplete, inaccurate, or missing information.

Today's fashion industry-focused PLM market is dominated by a number of global key players, including APTOS, Backbone, Bamboo Rose, CBX, Centric Software, CGS, Coats Digital–VisionPLM, Delogue, Infor, Lectra–Kubix Link & Gerber Yunique PLM, PTC–Flex PLM, and others. It is a growing market with an estimated value of US$171 million spent on new PLM projects in 2022 (Harrop, 2022a, p. 95).

11.5   COMPONENTS OF PLM SYSTEMS

PLM systems are not normally bought off the shelf ready-to-use because they come with a multitude of different options, each with its own list of modules, main processes, sub-processes, features, and functions which need to be configured to suit a specific business model. On average, there are more than 40–50 different processes included in a typical PLM solution, some of which are listed here, and each of which is likely to be of a different maturity level when compared to the industry average. Therefore, PLM systems are carefully tailored by the PLM vendor to meet the individual requirements of each business user because every business will have different ways of managing the product lifecycle.

Examples of typical components or modules found in a retail, footwear, and apparel PLM solution include:

- PLM Libraries (seasons, measurements, trims, components, materials, suppliers, product types, templates, critical paths, colours, etc.)
- Story and Mood Boards
- Digital Asset Management (DAM)
- NPI (New Product Introduction)
- BOM (Bill of Materials)
- BOL (Bill of Labour)
- Costing
- Sourcing
- Document and File Management
- Release or Change Management
- Access or Organisation Management
- Product Configuration and Template Management
- Critical Path and Lifecycle Management
- Bill of Process: how we process materials, garments

(Harrop, 2022b)

A PLM system should also offer integration with common systems like 2D CAD, 3D CAD, ERP, and other best-of-breed business systems as this enables brands and manufacturers to operate seamlessly and efficiently. Most PLM software today can be integrated with almost all business systems, including ERP, CRM, e-Commerce, Planning, Storyboards, Microsoft Office, social media (Instagram, Slack etc.), Adobe Creative Suite, 3D, Material Platforms, 2D CAD/CAM, Sourcing Platforms, Labour Costing, Machine IoT connections, 3D virtual simulation, and many more. Beyond these integrations, Harrop (2022a) argues that PLM is the data repository, and what we are seeing today is a set of new technology ecosystems the likes of downstream marketing and e-commerce, upstream digital product creation (3D), upstream manufacturing and collaboration (Tech Packs), financials, and logistics. Each of these ecosystems will operate in a closed loop but will interface seamlessly into the PLM backbone to send and receive data.

11.6 ADVANCES IN FASHION INDUSTRY PLM SYSTEMS

A variety of factors are currently driving advances in PLM systems. They include the post-pandemic working practices associated with the accelerated pace of digitalisation in the industry. Also, the post-pandemic consumer is assisting to drive the industry towards transparent circular supply chain practices that rise above greenwashing in meeting and communicating environmental due diligence and social compliance. These factors are exerting pressure on fashion businesses and manufacturers to compile detailed labour and wage data in the form of a Bill of Labour (BOL) as part of an overarching Bill of Process (BOP) that only the sophistication of PLM systems can manage.

Other advances include SaaS-licensing solutions that further support interoperability across the value chain, many of which are currently design-focused integrations, for example in the form of two-way connectors that link PLM to DPC-3D ecosystems that support design assets generated with 3D fashion design software such as CLO3D, Browzwear V Sticher, or Tuka3D, as well as other plugins that enable the two-way communication between other notable software solutions used by the fashion industry. Advances in PLM systems are also supporting further growth in the PLM market, and this is having a positive impact in terms of the development of new fashion industry-focused PLM jobs and careers that were unimaginable just a few years ago.

All of these PLM advances are summarised here and will be considered in further detail in the rest of this section:

- Post-pandemic working practices/post-pandemic consumer
- Sustainability and transparency for social compliance (using scientific impact calculations)
- Bill of Labour
- Interoperability: design-focused integrations and plug-in technologies
- Licensing solutions that support collaboration with partners in the value chain

11.6.1 Post-pandemic Working Practices and the Post-pandemic Consumer

Emerging out of the pandemic era it has become evident that the way the fashion industry works has changed. The industry is becoming increasingly digitalised, and fashion brands are buying smaller more complex orders driven via near real-time consumer demand. The new consumer not only expects faster fashion with smaller, more personalised collections more frequently, but they also expect greater transparency and sustainability across the value chain. This is also supporting a movement towards non-seasonal slow fashion and circular supply chain models. According to PLM Report (2022c), PLM can play an essential role in managing circular models whereby the processes of rent, repair, refurbish, and recycle become part of the value chain. This is already happening as the leading brand Zara launched a resell, repair, donate campaign on 4 November 2022 to enable their consumers to extend the life of clothing purchased from their brand (Independent, 2022).

For example, customers can request a repair such as a seam repair, the addition of a button, or de-bobbling. The customer can book the repair online and pay a standard fee for each type of repair. Being able to link a pre-owned item back to a style within a PLM system can make the tracking of fabric and trims needed for repair, a simple and quick product code search.

At the same time, to reduce waste associated with overproduction and markdown of apparel due to the complexities associated with apparel fit, opportunities to support the demand for made-to-measure apparel can be effectively managed using a PLM system. For example, the Gerber Accumark Made to Measure (MTM) software is linked to Gerber's Yunique PLM system and enables the creation of a customisable database with smart search tools for querying orders by customer name or other preferred detail. The software also reduces time to market by automating order selection and decision-making using powerful knowledge-based rules. Other automatic features include the potential to further streamline productivity and maximise material utilisation through the ability to generate and export cut data via its integration with AccuNest™ which means that markers can also be generated automatically.

Surviving and prospering in the post-pandemic digital future will require businesses to launch new style options, services, and varieties at a much faster pace and produce them leaner and to a higher quality. According to Harrop (2022a), this cannot be achieved without the use of an integrated, multifunctional, enterprise-wide PLM solution (Conlon, 2020; Harrop, 2022a; Lunoe, 2022; McCready-Stocks, 2022).

11.6.2 Sustainability and Social Compliance

With increasing sustainability and social compliance regulations coming into force across many nations, along with growing consumer demand for proof of environmental and social improvements in value chains, it is becoming necessary for fashion brands to disclose their environmental and social impact right the way through their value chain whilst meeting new and emerging regulatory and consumer-driven requirements which fall into two key areas:

- Due diligence and transparency
- Marketing and product claims

Laws governing due diligence have already been passed in some countries mandating that companies have visibility over their supply chains. This means that companies are responsible for identifying if any of their suppliers are at risk and to resolve this with their supplier by implementing policies that mitigate such risk whilst implementing any necessary solutions. There are also a number of mandatory regulations that have been passed by governments in the United States, Germany, France, and Norway, such as the US Uyghur Forced Labour Prevention Act, New York Fashion Sustainability and Social Accountability Act, German Supply Chain Due Diligence Act (LkSG), French Duty of Care Law, Norwegian Transparency Act. To date, the UK government has not implemented any mandatory legislation surrounding Human Rights and Environmental Due Diligence (HREDD); however, according to BHRRC (2022), in September 2022, over 47 major UK companies, investors, and business associations, which include John Lewis, Tesco, ASOS, Primark, Unilever, The British Retail Consortium and many others, have called on the UK government to introduce a HREDD law that mandates companies to carry out human rights and environmental due diligence. To demonstrate HREDD compliance, most current legislation requires companies to be able to make available for each product, evidence such as:

- Certification of origin
- Detailed evidence of source of merchandise and labour used in production
- A detailed chain of custody from the bale to final production including all parties and suppliers involved

It is evident that it is not possible to capture or control this volume of data without a dedicated traceability system, and this has led to the development of advanced supply chain traceability software companies such as Trustrace, whose highly sophisticated software uses AI to streamline data collection which can be plugged into existing PLM systems via openAPI (Turstrace, 2022).

In terms of effectively managing marketing and product claims associated with sustainability and social compliance, this process is experiencing ever-increasing scrutiny as concerns continue to mount regarding the level of inaccurate information being communicated to consumers. According to Chan (2022), a recent study has found that nearly 60% of green claims made by 12 major fashion brands in the UK and Europe were unsubstantiated or misleading. Fashion brands are therefore under pressure to eliminate greenwashing by communicating accurate and easy-to-understand information relating to the environmental and social impact of their products.

In terms of effectively managing marketing and product claims associated with sustainability and social compliance, fashion brands are under pressure to eliminate greenwashing by communicating accurate and easy-to-understand information relating to the environmental and social impact of their products. According to Chan (2022) greenwashing has grown to such an extent that a recent study published by the Changing Markets Foundation in 2021 found that nearly 60% of green claims made by 12 major fashion brands in the UK and Europe were unsubstantiated or misleading. A recent example that demonstrates just how challenging this process can be is The Higg Index, developed by leading fashion brands including Walmart, H&M, Nike, Levi's, and Patagonia and managed by the Sustainable Apparel Coalition (SAC) and their technology partner Higg. The Higg Index was designed to assess the sustainability of materials used in fashion products with integrated links to many of the world's leading fashion PLM platforms. In 2021, a consumer-facing section of the Higg Index was developed to enable consumers to view the environmental impact of individual materials. It was suspended in June 2022 because it was found to be inaccurate, favouring synthetic fabrics as more sustainable than natural fabrics (Shendruk, 2022).

Moving forward, it is now possible to scientifically measure greenhouse gases (GHG) on a per metre basis and to calculate CO_2 impact measures at a garment level for the first time. These developments can positively support product sustainability and social compliance and therefore work towards the elimination of greenwashing by enabling the tracking and logging of materials using the PLM Bill of Process right down to composition level to make sure styles are designed to live longer, with the ability to rent, repair, refurbish, resell, and recycle.

There are already several companies, like Made2Flow, who have partnered with PLM leaders such as Infor to increase transparency in the production supply chain using modelisation engines to build a process list for brands, retailers, and manufacturers and measure environmental indicators such as GHG, water depletion, and land use that can all be linked back to choices and assessment methodology (Forrest, 2022).

Another near-future development is the potential for product passports where data for a product will be stored and shared with consumers. This goes beyond traditional PLM software to cover the end-to-end sustainable lifecycle of a product, where PLM is a key part of the total solution (Amed et al., 2021; Harrop, 2022a).

11.6.3 BILL OF LABOUR

The BOL is not a new PLM module; however, it is regaining prominence as an important tool for tracking and meeting social compliance for workers employed not only by main suppliers but also by the sub-sub-supplier as part of the tiered supply chain.

This data collection cannot easily be generated or stored without a dedicated PLM platform that can enable all suppliers and manufacturers to systematically record and share data that can be used externally for transparency to consumers, stakeholders, and for governmental reporting requirements, which may become necessary in the near future (PLM Report, 2022c).

Several of the leading fashion industry PLM companies such as Infor PLM and Coats Digital VisionPLM support complex BOL specification involving data such as standard minute values and a myriad of details concerning the labour associated with production operations.

In addition to achieving social compliance, rising labour costs around the world are also driving renewed interest in BOL, and this is necessitating brands and manufacturers to invest in the development of more accurate and transparent costings. In response to this Coats Digital has recently launched GSDCost. GSD™ was originally established in the 1980s and was used by brands and manufacturers to optimise time–cost benchmarks for garment costing and manufacturing excellence.

GSDCost works by analysing all direct labour activity from cutting to the packing of finished goods and allocates each element of labour activity a predetermined standard time-in-motion code.

These codes can then be assembled in multiple combinations that describe and benchmark the manufacturing method and associated labour costs during planning. It also enables cost engineering to ensure garments are designed and developed to the right price point thereby optimising margin and fact-based costing. Recognised by the International Labour Organization (ILO), it is the only globally recognised BOL tool which combines the international standard time for any given style, with detailed factory efficiencies, contracted hours, and the fair living wage for the country, independently provided by the Fair Wage Network. This allows brands and retailers to quickly agree the fair living wage allowance for any given garment, in any factory in the world.

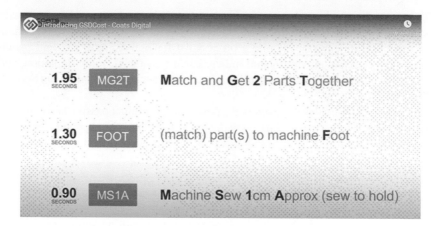

FIGURE 11.1　GSDCost allocation of predetermined time and motion codes.

Source: Courtesy of Coats Digital, www.CoatsDigital.com

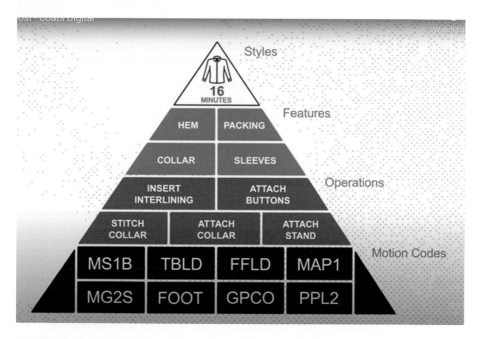

FIGURE 11.2　GSDCost time and motion codes for manufacturing method and labour costs.

Source: Courtesy of Coats Digital, www.CoatsDigital.com

Looking further forward, according to Harrop (2022a), over the next few years, we can expect to see greater adoption and acceptance of blockchain for the tracking and tracing of raw materials all the way from the factory to the consumer. Digital certification supported on the blockchain will start to cross the chasm from proof-of-concept into the mainstream, supporting proof of provenance linked to real-time transparency. Ultimately, this will give the consumer trust in who really made their product (Harrop, 2022a).

11.6.4 Design-Focused Integrations and Plugins

For retailers, brands, and manufacturers to operate seamlessly and efficiently, the software that they utilise must be able to talk to each other, and both design-focused integrations and plugins are supporting designers to spend more time designing and less time administering data versions.

These design-focused integrations and plugins are in essence a software add-on that is installed on a program such as PLM to enhance its capabilities. The first PLM companies to develop 3D plugins included the collaboration between Freeborders and Browzwear in the mid-2000s and also Gerber technology who integrated its own PLM solution to its proprietary three-day solution circa 2015 along with Lectra fashion PLM who also integrated its PLM solution to its proprietary 3D solution circa 2017.

More recently, in 2018, Centric PLM has launched a 3D sample review mobile app to review digital samples in virtual fit scenarios within its Centric PLM system. The mobile app enables users to review digital samples created by 3D vendors or in-house artists in virtual fit scenarios. Users can add comments and receive amended digital samples directly within Centric PLM. These digital assets can be attached to style data and incorporated into active line plans, giving an up-to-date view of where each product is in the sampling and review process.

In 2019, Centric PLM developed a fully digital, 3D design and development work-flow with integration to the leading 3D fashion design software solutions such as CLO, Browzwear, and EFI Optitex, providing a seamless 3D workflow using 3D information for ideation, design and fit reviews, tech pack development, requests for quotation, and co-creation with suppliers for effective and fully inclusive digital transformation around the supply chain (Harrop, 2022a). According to Centric (2019), its 3D Connectors offer benefits such as reduced time to market, lowered development costs, increased product innovation, and a smaller environmental footprint. Centric believes its connectors also drive efficiency for fashion and apparel brands specialising in personalisation or made-to-measure by automatically updating product designs and fit via virtual samples. They can also share photorealistic 3D models with customers for review and approval prior to production.

Another example of a design-focused integration is the CLO-Vise Plugin, developed by GoVise Technologies and CLO3D (GoVise, 2022). PTC, who is a market leader in fashion industry-focused PLM systems, was the first PLM company to partner with CLO3D to integrate this plugin into its FlexPLM system. This bidirectional integration enables CLO3D fashion design software to be integrated with PLM systems, so that designers can either push data from CLO3D to the PLM or retrieve PLM data into CLO and publish the garment, BOM, or tech pack and send them back to PLM when it is ready for review with cross-functional teams. This means designers can quickly access 3D assets and spend more time creating new designs and less time administering and uploading data into PLM software.

Other benefits for designers include access to product data such as colour, trim, and material libraries from Flex PLM whilst working within CLO3D via the CLO-Vise Plugin. According to JustStyle (2020), this plugin can assist fashion brands to improve efficiencies by reducing time, development costs, and material consumption. Once again, this means more time focused on designing rather than on administration tasks.

Since 2020, the CLO-Vise plugin has proven highly successful and several other leading fashion industry PLM companies such as Bamboo Rose have partnered with 3D fashion design software companies such as CLO3D in 2022 to customise the CLO-Vise plugin for their PLM system that they refer to as a 3D Smart Connector. Like the PTC CLO-Vise plugin, the Smart Connector allows designers to easily access and reuse 3D digital assets including colour and materials libraries within the Bamboo Rose PLM. According to Cison (2022), this integration enables Bamboo PLM users to view a spinnable 3D design image that is attached to the product record.

Even though the use of 3D fashion design software continues to accelerate, the use of 2D design software, and in particular Adobe Illustrator, continues to flourish. PTC pioneered the first bidirectional integration between FlexPLM and Adobe Illustrator in 2012 when other vendors simply dismissed it and didn't see that designers needed to tap into PLM functionality. PTC has continued to make significant updates to this plugin in response to the growing need for design–development–efficiency.

The plugin makes it easy to share sketches between Adobe Illustrator and PTC's FlexPLM using familiar functions such as drag-and-drop, enabling designers to synchronise design assets such as technical sketches, colour-ups, and range boards from Adobe Illustrator directly to PLM. This eliminates the need for manual uploads and also enables technical and production teams, factories, vendors, buyers, merchandisers, and retail teams to collaborate on new designs in real time. This industry standard connectivity eliminates many adoption barriers that creative teams face when using typical PLM interfaces.

11.6.5 LICENSING SOLUTIONS THAT SUPPORT COLLABORATION WITH PARTNERS IN THE VALUE CHAIN

Having emerged from the recent global pandemic, the way that we work and collaborate has been transformed by digital technologies. According to Willis (2022), in the past, PLM was simply a place in which to store product information. This has changed due to the increased variety of users and variables which require greater collaboration between the design team, the merchandising team, the QA team, corporate social responsibility (CSR), and sustainability teams but also crucially with all partners involved in the value chain so everyone can benefit from rich data and timely updates. According to PLM Report (2022b), connecting all of your suppliers to your PLM system is imperative to not only achieve speed and efficiency but even more compelling for achieving end-to-end visibility and traceability.

PLM systems have been traditionally expensive to implement, and many external partners in the value chain have remained locked out of the comprehensive data that PLM systems manage; however, this is beginning to change. PLM leader Delogue PLM has developed a two-way supplier collaboration functionality that enables internal users to collaborate with all external users at every stage of engagement, offering accountability and transparency within their digital inventory. This provides an environment of automation with less likelihood of errors that trickle downstream. For instance, its care instructions module invites suppliers to engage with brands in the input of accurate data through predefined fields. This helps to assure that the product is true to the information given, and this ultimately leads to increased consumer satisfaction and brand trust. To assure supplier compliance and engagement with the PLM, Delogue PLM is currently the only PLM provider who doesn't charge suppliers a licensing fee. This is because it has proven that the quality and accuracy of product data improve when as many stakeholders from the value chain collaborate because brands are becoming increasingly reliant on suppliers providing valuable information. Recent examples include mandatory EU directives around human rights and environmental due diligence which will potentially become more regulated in the future.

11.7 FUTURE ADVANCES IN PLM TECHNOLOGY

In the last couple of years many fashion businesses have moved forward with progressive digitalisation strategies supported by powerful PLM platforms that can extend the reach of PLM to upstream manufacturing processes in the pursuit of digitally linking factory machinery and processes that enable the sharing of real-time production data that in turn can be used to support real-time transparency and monitoring.

To achieve this next level of interoperability, manufacturers will need to first deploy their own internal PLM solutions to their internal hardware and software and then utilise the internet of things (IoT) devices to connect into their hardware machines. To maximise the benefits, these digitally connected factory platforms will need to be integrated with a range of external systems involving not just retailers or brands, such as PLM Tech-Pack or Enterprise Resource Planning contracts and Purchase Order solutions but also to a range of third-party best-of-breed infrastructure that are ideally sharing a common data set, driven via a shared retail and factory workflow and critical path tasks.

There are already several leading fashion groups and manufacturers that have begun to embrace extended value chain platforms utilising the IoT, such as sensors located within critical factory hardware and software solutions, including factory planning that involves 2D CAD and CAM, testing, spreading, numerically controlled cutting, costing, sewing, pressing, quality assurance, and control and packaging, and so on. These connections can each feed real-time performance data into a shared digital platform(s) that are then able to analyse the resulting data-monitoring during each step of the process. They can also be supported by smart algorithms together with AI and machine learning (ML) applications, sharing real-time results, insights, and predictions between each of the value chain partners, ultimately combining to make better operational decisions. The analysed-factory operational information can also be used in identifying opportunities for the factory management team to help improve factory efficiency, down to the individual machine or operator level, including efficiency and performances, optimising factory throughputs, identifying bottlenecks early in the process, product quality related to an individual machinist or machine issues, resulting in limiting defects, reducing product returns, amongst many of the benefits coming from shared real-time data.

It is obvious from the latter that the capabilities of PLM have changed dramatically in a relatively short period of time. Not so long ago PLM was referred to as the exclusive backbone of a company; however, this perspective has little relevance to fashion businesses that are successfully progressing their digitalisation strategy. These PLM users are experiencing PLM for what has become which Harrop (2022a) describes as 'multiple, interconnected ecosystems that are starting to make up foundational technology stacks' (Harrop, 2022a, p. 100).

As the uptake and use of PLM amongst retailers and brands continue to increase, new skill requirements and new careers are emerging including the roles of PLM experts or digital transformation consultants whose digitalisation projects are projected to support further expansions of PLM access to new categories of users and to new stakeholders in the value chain.

As the fashion industry starts to implement its digital factory-connected strategy, just like retailers and brands before, they will need to carefully consider upskilling their digital workforce and will need to consider employing or developing their own technology expertise, which will require the upskilling of a new digitally savvy workforce that will be required to deliver the level of change required for the future. The need to protect PLM systems and their stakeholders will also become more critical in the near future meaning 'cyber talent will be at a premium' (Amed et al., 2021, p. 2).

Both the fashion industry and higher education have an important collaborative role to play in developing the workforce requirements of the increasingly digitalised fashion industry. In response to this, PLM vendors such as Delogue PLM and Coats VisionPLM have developed educational programmes to support higher education institutions to prepare future-ready fashion professionals for the fashion industry where PLM is now recognised as a necessary strategic tool for digital transformation (Conlon, 2020; Harrop, 2022a; Lunoe, 2022; McCready-Stocks, 2022).

REFERENCES

Amed, I, Berg, ABA, Hedrich, S, Merle, JEJLL and Rölkens, F (2021) *State of Fashion 2022: An Uneven Recovery and New Frontiers*, McKinsey & Company [online], www.mckinsey.com/~/media/mckinsey/industries/retail/our%20insights/state%20of%20fashion/2022/the-state-of-fashion-2022.pdf (Accessed 02/01/2024)

Backbone (2022) *Backbone PLM Unveils the New Adobe Illustrator Plugin* [online], https://bamboorose.com/blog/backbone-plm-enhances-design-workflows-with-adobe-illustrator-plugin/ (Accessed 02/01/2024)

BHRRC (2022) *UK: Businesses and Investors Call for New Human Rights Due Diligence Law, Business and Human Rights Resource Centre* [online], www.business-humanrights.org/en/latest-news/uk-businesses-and-investors-call-for-new-human-rights-due-diligence-law/ (Accessed 02/01/2024)

Centric (2019) *Centric Software PLM Inks Partnerships with CLO, Browzwear and EFI Optitex* [online], www.centricsoftware.com/press-releases/centric-software-plm-inks-partnerships-with-clo-browzwear-and-efi-optitex/ (Accessed 02/01/2024)

Chan, E (2022) *Can a New Site Help Tackle Greenwashing in Fashion?* [online], www.vogue. co.uk/fashion/article/greenwashing-fashion-website (Accessed 02/01/2024)

Cison (2022) *Bamboo Rose Launches 3D Smart Connector with CLO Virtual Fashion to Reduce Sample Cycles and Speed Time to Market* [online], www.prnewswire.com/news-releases/ bamboo-rose-launches-3d-smart-connector-with-clo-virtual-fashion-to-reduce-sample-cycles-and-speed-time-to-market-301636085.html (Accessed 02/01/2024)

Conlon, J (2020) *From PLM 1.0 to PLM 2.0: The Evolving Role of Product Lifecycle Management (PLM) in the Textile and Apparel Industries* [online], www-emerald-com. mmu.idm.oclc.org/insight/content/doi/10.1108/JFMM-12-2017-0143/full/ (Accessed 28/12/2023

Corallo, A, Latino, ME, Lettera, S, Marra, M and Verardi, S (2013) Defining Product Lifecycle Management: A Journey across Features, Definitions and Concept. *ISRN Industrial Engineering*: 1–10 [online], www.researchgate.net/publication/258390180_Defining_ Product_Lifecycle_Management_A_Journey_across_Features_Definitions_and_ Concepts (Accessed 03/01/2024)

Forrest, F (2022) *Made to Flow Team on Transparency* [online], www.just-style.com/news/ infor-made2flow-team-on-transparency/'?cf-view (Accessed 02/01/2024)

GoVise (2022) *Fashion 3D Integrations* [online], https://govisetech.com/product-offering/ (Accessed 03/01/2024)

Harrop, M (2022a) PLM Report 2022, *The Interline* [online], www.whichplm.com/pub-lication/plm-buyers-guide-2022/#:~:text=The%20PLM%20Report%202022%20 is,PLM)%20content%20to%20The%20Interline (Accessed 03/01/2024)

Harrop, M (2022b) *The Bill of Process* [online], www.theinterline.com/2021/09/15/the-bill-of-process-bop/ (Accessed 03/01/2024)

Independent (2022) *Zara to Launch Pre-Owned Service for Shoppers to Resell, Repair or Donate Items* [online], www.independent.co.uk/life-style/fashion/zara-pre-owned-clothing-resell-b2207613.html (Accessed 03/01/2024)

The Interline (2021) Interline Identifies: Amir Lehr, CEO Optitex, *The Interline* [online], www.theinterline.com/2021/02/24/interline-identities-amir-lehr-ceo-optitex/ (Accessed 30/12/2023)

Luciano, C (2010) *Manufacturing Pioneers Reduce Costs by Integrating PLM and ERP* [online], https://community.dynamics.com/blogs/post/?postid=a2f8b66e-0261-457e-9ec4-0169af3f7231 (Accessed 03/01/2024)

Lunoe, J (2022) Delogue PLM, cited in PLM Report 2022, *The Interline* (p. 68) [online], www.whichplm.com/publication/plm-buyers-guide-2022/#:~:text=The%20PLM %20Report%202022%20is,PLM)%20content%20to%20The%20Interline (Accessed 03/01/2024)

McCormick, H, Cartwright, J, Perry, P, Barnes, L, Lynch, S and Ball, G (2014) Fashion Retailing, Past, Present and Future. *Textile Progress*, 46(3) [online], www.tandfon-line.com/doi/citedby/10.1080/00405167.2014.973247?scroll=top&needAccess=true (Accessed 28/12/2023)

McCready-Stocks, S (2022) cited in PLM Report 2022, *The Interline* (p. 62) [online], www.which-plm.com/publication/plm-buyers-guide-2022/#:~:text=The%20PLM%20Report%20 2022%20is,PLM)%20content%20to%20The%20Interline (Accessed 03/01/2024)

PLM Report (2022a) *In Converstion with Matthew Bonenfant VP Marketing Strategy Fashion*, pp. 77–79 [online], https://theinterline.com/PLM-Report-2022.pdf (Accessed 02/01/2024)

PLM Report (2022b) *In Conversation with Fabrice Canonage* [online], https://theinterline. com/PLM-Report-2022.pdf (Accessed 02/01/2024)

PLM Report (2022c) *In Conversation with Andrew Dalziel VP Industry and Solutions Strategy* [online], https://theinterline.com/PLM-Report-2022.pdf (Accessed 02/01/2024)

SAP (2022) *What Is Product Lifecycle Management* [online], www.sap.com/uk/products/scm/plm-r-d-engineering/what-is-product-lifecycle-management.html#:~:text=A%20PLM%20software%20system%20is,process%20for%20all%20business%20stakeholders (Accessed 02/01/2024)

Segonds, F, et al. (2014) Early Stages of Apparel Design: How to Define Collaborative Needs for PLM and Fashion? *International Journal of Fashion Design, Technology and Education*, 7(2): 105–114 [online], www.tandfonline.com/doi/full/10.1080/17543266.2014.893591 (Accessed 02/01/2024)

Shendruk, A (2022) *The Controversial Way Fashion Brands Gauge Sustainability Is Being Suspended* [online], https://qz.com/2180322/the-controversial-higg-sustainability-index-is-being-suspended (Accessed 02/01/2024)

Turstrace (2022) *Turstrace Supply Chain Traceability Software*, TrusTrace—Leading Fashion Supply Chain Traceability Software [online] https://trustrace.com/

Willis, R (2022) Aptos PLM cited in PLM Report 2022, *The Interline* (p. 67) [online], www.whichplm.com/publication/plm-buyers-guide-2022/#:~:text=The%20PLM%20Report%202022%20is,PLM)%20content%20to%20The%20Interline (Accessed 02/01/2024)

Zaczkiewiez, A (2021) *PTC's Flex PLM Now Enables Automatic Optimization of 3D Assets*, [online], https://wwd.com/business-news/technology/ptc-flexplm-upgrade-1234799928/ (Accessed 02/01/2024)

12 Advances in Sewn Product Technology for Tomorrow's Jobs

12.1 INTRODUCTION

This chapter will focus on how advances in sewn product technology have disrupted and informed the development of new skills, knowledge, and jobs during each consecutive Industrial Revolution, and how this can directly inform the skills and knowledge required for current and future employment in the fashion industry. The chapter will begin by considering how the advances in sewn product technologies that were introduced in the first Industrial Revolution enabled the apparel industry to develop and take shape whilst advances introduced in the Second Industrial Revolution enabled it to dramatically expand and prosper. Consideration will be given to how the industry's skills, knowledge, and job requirements were facilitated during this expansive period. The chapter will move on to consider the knowledge, skills, and jobs that were developed in response to the Third Industrial Revolution and both their present relevance and the need for their vital transformation in light of the latest advances in sewn product technology that are now driving the Fourth Industrial Revolution. This chapter will conclude by outlining some of the new and emerging fashion industry jobs, including the knowledge and skills base they require and what can be done to support a relevant employment stream for the fashion industry of the future.

12.2 SKILLS TO SUPPORT ADVANCES IN SEWN PRODUCT TECHNOLOGY FOR THE FIRST INDUSTRIAL REVOLUTION (1IR): 1760–1860S

According to Dean (2000), the first Industrial Revolution, also known as 1IR, began in Great Britain between the period 1760 and 1860, and is described as the starting point of a global shift away from hand-made craft-based fabrication towards mechanisation and large-scale factory production.

The conditions that assisted to bring about the 1IR are summarised by Michon (2022), who argues by the mid-eighteenth century, Britain had become a colonial power *with access to* abundant raw materials such as cotton. Also, it had a wide trade network and a shipping industry to move raw materials and finished goods around the world, as well as energy reserves including coal and water power and a growing population to serve as workforce. The British Industrial Revolution started in the textile industry and was enabled by a string of important textile-related inventions such as the flying shuttle and the spinning jenny, which were supported by other

DOI: 10.1201/9781003126454-12

key inventions that assisted in connecting them to a power source such as Richard Arkwright's Waterframe, which in turn enabled the development of one of the first textile factories employing over a thousand people at Cromford in Derbyshire. Other advancements that further supported industrialisation included the development of the lockstitch machine in the 1840s which in turn facilitated the development of the apparel industry. Likewise, the introduction of the steam-powered engines adopted by the textile, coal, and iron industries led to the development of steam-driven locomotives in 1820. The 1IR led to improvements in the standards of living, health and prosperity, and education for most of the population. Dean (2000) argues that its history is of special interest to modern economic development research because it enabled the social mobility of those previously in poverty, which still has relevance to some of today's developing countries.

In terms of individuals accessing education and training to develop the technical skills to facilitate the 1IR, according to (Howes, 2017) during this time, Britain established a successful apprenticeship system. Also, it had a well-educated and highly skilled elite, many of whom possessed what he calls an *improving mentality*, a mindset for seeing opportunities for improvement to any industry they believed could and should be enhanced by further self-education as well as collaborating with the expertise of others.

12.3   SKILLS TO SUPPORT ADVANCES IN SEWN PRODUCT TECHNOLOGY FOR THE SECOND INDUSTRIAL REVOLUTION (2IR): 1870–1950S

The Second Industrial Revolution also known as the technological revolution or 2IR was a period of groundbreaking advancements in technology, manufacturing, and industrialisation between the period 1870 and 1950, which led to the development of new industrial sectors such as petroleum, chemical, steel, business management, apparel, and many others. It initially developed in Britain straight after the 1IR and spread to the rest of Europe and America. It was known as the technological revolution because it involved the progressive development of groundbreaking advancements in technologies and manufacturing such as the introduction of electrification to replace water, wind, and steam power, the introduction of Fordism and Taylorism enabling the development of assembly lines and the division of labour for mass production. Innovations in communication including the telegraph, telephone, and radio as well as the introduction of new modes of transport including canals, railways, automobiles, and aeroplanes not only supported industrial progress, but it also enabled the movement of people around the world such as the UK's migration policy, which aimed to attract migrant workforce to rebuild the economy weakened by the war years and provide a labour force to work in UK's manufacturing sector that rapidly developed after the 1950s (Windrush Team, 2022).

The apparel industry rapidly developed at the beginning of the 2IR and was supported by the widespread availability of cloth enabled by ongoing advancements in the textile industry. Its development was further supported by the electrification of the industrial sewing machine and the new organisation and management principles of Taylorism and Fordism that had been developed within associated industries. The ability to mass produce apparel dramatically transformed the production of civilian clothing and military uniform manufacture which in turn stimulated further advancements and innovations in the quest for improved quality and efficiencies.

The fashion industry that developed in the UK during the 2IR was an industry based on analogue and mechanical methods. Large UK factories developed in response to the growing post war populations and utilised local labour who needed employment. Designs, patterns, and even markers for bulk production were constructed by hand, and all industrial sewing operations were labour-intensive with some simple forms of sewing automation emerging by the 1950s such as semi-automated sequential button sew machines. Business communication was supported by physical meetings or telephone and fax, and all documents including schedules, planning, and orders were either written by hand or typed on a manual typewriter as computers, the internet, and emails were not available at this time. Such business methods continued until personal computers became affordable and widely available along with the advent of emails in the mid-1990s. The key job roles that existed in the fashion industry at this time included designer, design studio manager, pattern cutter, sample machinist, marker maker, lay-planner, garment technologist, quality control, fabric technologist, buyer, marketing, cutting room manager, machine engineer, factory manager, accountant, warehouse manager, merchandisers. The duties undertaken in each of these roles varied from business to business; for example merchandisers were often involved in work associated with the bulk ordering of materials and trims but also the transportation of garments from factory to warehouses and the logistics of moving orders from warehouse to retail, and part of this role turned into a new stand-alone role called fashion logistics.

In terms of the development of skills and knowledge to support professional occupations in 2IR, prior to the development of widespread higher education in the UK in the 1990s, and depending on the nature of the subject, training for design-based subjects was undertaken at art schools some of which became fashion schools after the 1930s. Business management was predominantly taught within business schools at select universities, such as the University of Birmingham, which opened the first business school in the UK in 1902. Technical subjects were taught at mechanical institutes or technical colleges, most of which offered day release or evening courses to support work-based learning or self-development. As the twentieth century progressed, many of these courses were delivered by polytechnics originally set up in the 1960s with a focus on STEM subjects until they became independent universities with their own degree-awarding powers because of the Further and Higher Education Act 1992 (TEM, 2022).

12.4 SKILLS TO SUPPORT ADVANCES IN SEWN PRODUCT TECHNOLOGY FOR THE THIRD INDUSTRIAL REVOLUTION (3IR): 1950S–2008

According to Von Scheel (2022), the Third Industrial Revolution, also known as 3IR, began in the United States in the mid-1950s with ongoing new developments in electronics robotics and information technology that assisted in automating mass production (Von Scheel, 2022).

Also known as the Digital Revolution, it was a period of groundbreaking advancements in digital technology between the period 1950 and 2008 and was characterised by the progressive movement from analogue to digital from the mid-1980s onwards. Early examples of this change within the apparel industry include the introduction of CAD software such as Adobe Illustrator for apparel design or the use of CAD for the creation of digital patterns using software such as Gerber's AccuMark or Lectra Modaris, as well as the use of CAD and CNC to digitally control machining tools such as 3D printers, automatic cutters, and spreaders. Advances in CNC technologies during the 1980s also enabled the development of the first semi-automatic sewing machines assisted by specialised sewing clamps and jigs (Hayes and McLoughlin, 2008).

However, competition from cheap imports and rising labour costs meant it was no longer cost-effective to manufacture apparel in the UK. This was compounded by the introduction of the Minimum Wage Act in 1998. Whilst some industries such as the automotive industry invested in automation to reduce manufacturing costs in the UK, the apparel industry was forced to offshore the labour-intensive work of garment making to include spreading, cutting, and garment making to low-wage countries because of the complexities associated with fully automating the handling of non-standard materials (Tyler, 2009). In an attempt to increase efficiencies further, many UK-based apparel enterprises also offshored professional occupations, including sample room manager, sample machinist, pattern cutter, lay-planner and marker maker, garment technologist, quality control, fabric technologist, cutting room manager, factory manager, machine engineer, warehouse manager. The professional occupations that remained in the UK included those associated with design, buying, merchandising/import merchandising, allocation, marketing/e-commerce marketing, e-commerce developer, analytics, and logistics. Jobs such as warehouse managers remained as warehouses were needed at both ends of the manufacturing supply chain.

During this time, many of the jobs that were offshored involved the upskilling of local employees with expat employees as the trainers. It is notable that many of them became redundant once training and development of the new workforce were complete. Although labour in offshore locations was cheap during this period, many of the semi-automated machines that were developed in the 1980s for the purpose of protecting manufacturing in the West were also offshored to further enhance quality and efficiencies but were eventually needed to offset rising labour costs in response to economic development facilitated by the industrialisation of the developing nations.

By the millennium, major advancements in computing power and GPU technology facilitated advancements in 3D apparel prototyping that were further integrated with other digital technologies to include 3D body scanning, 3D knitting, laser cutting, fully automated cutting and spreading technologies, and fully automated sports shoe manufacture. Other more recent advancements have included the introduction and ongoing research of cobots and sewbots to enable the fully automated manufacture of apparel to facilitate nearshoring and onshoring. To date, uptake of 3D prototyping has been slow but is gathering pace, and both use and uptake of cobots in the fashion industry are currently extremely limited but forecasted to grow rapidly.

Sources of education to support skill and knowledge acquisition in response to this revolution have been similar to the 2IR. In terms of the needs of fashion businesses who need to upskill, this has often involved in-house training where experienced staff train others or through the use of specialist training providers who deliver bespoke training courses for their specific hardware or software. This is often very expensive and therefore limited as an approach for smaller businesses. Poaching trained personnel from competitor organisations has persisted as an iniquitous recruitment strategy in most businesses including the fashion industry throughout 3IR but has limited relevance to 4IR as talent pools are only establishing themselves in response to new 4IR technologies (Honeyman, 2000; McDowell, 2021). Other recruitment strategies employed during 3IR have included employing temporary specialist freelancers many of whom are self-taught or have transferred from related industries (Roberts-Islam, 2020). This approach has allowed permanent employees time to upskill. Towards the latter part of 3IR, many individuals have been able to self-teach advanced software due to the widespread availability of free or affordable online learning courses both during and subsequent to the pandemic which according to McKinsey (2022, p. 3) 'assisted to push the need for a new learning experience online'.

Art schools and both pre and post UK HEI's have also played a major role in educating the professional workforce for 3IR. As previously mentioned, design-based subjects were until the 1990s most often delivered by fashion schools, many of which sat within UK Art Schools such as Manchester Polytechnic's School of Art or the Royal College of Art (RCA). Business management continued to be taught within business schools which sat within most of the major universities in the UK and could often include aspects of buying, merchandising, and logistics as opposed to specialist stand-alone courses in these areas. Marketing subjects often sat within economics as applied economics (Ferrell, 2015). Despite the offshoring of many technology-related jobs between the late 1980s and the millennium, demand for technology-based jobs in the UK apparel industry has continued to be sought after by retailers in response to the need to achieve high levels of quality and elevated efficiencies. Fashion technology subjects continued to be taught by polytechnics, which became independent universities with their own degree-awarding powers because of the Further and Higher Education Act 1992 (TEM, 2022). Most of these new universities have in recent years developed dedicated fashion courses to meet all the specialised careers that sit within the apparel/fashion supply chain.

The current challenge for many UK-based HEIs is to keep up with the pace of change whilst upskilling their employees and their curriculum, and some of the largest fashion schools such as Manchester Metropolitan University and the London College of Fashion are able to quickly forecast and deliver education solutions for new and emerging advances in sewn product technology.

12.5 SKILLS TO SUPPORT ADVANCES IN SEWN PRODUCT TECHNOLOGY FOR THE FOURTH INDUSTRIAL REVOLUTION (4IR): 2009–PRESENT AND BEYOND

The Fourth Industrial Revolution also known as 4IR began in Germany at the beginning of the millennium at a time when a minority of forward-thinking enterprises had already begun to embrace the move towards advanced digitalisation, advanced connectivity, and advanced automation. The term Industry 4.0 was used to describe this new industrial change within the title of a report presented to the German government in 2009 (Lasi et al., 2014). This point is supported by The Manufacturer (2019), which argues that the key trigger for the 4IR was Germany's national strategy in 2009 called the Digital Agenda. The Digital Agenda was a national masterplan involving industry, academia, and the government. Led by Henrik Von Scheel, a world expert in strategy and competitiveness, it was designed to overcome the economic crisis that was happening in Germany at that time. News of Germany's plan to launch a Fourth Industrial Revolution in May 2015 was under-reported in the media (Temperton, 2015) and in December 2015 appeared to be usurped by a journal paper called 'The Fourth Industrial Revolution', published within an influential foreign policy journal called *Foreign Affairs* by the leader of the World Economic Forum Klaus Schwab. This journal paper put forward a motivating and highly convincing case that the world was on the brink of an uncertain technological revolution that would change the way we live, work, and relate to one another (Schwab, 2015). Since then, Schwab has been instrumental and highly influential in the global dissemination and uptake of 4IR, which is currently being embraced by countries, such as Japan, China, the United States, Switzerland, Germany, Mexico and Columbia, and many others including Great Britain.

The 4IR is building on the 3IR, which was based on digitalisation and is characterised by a fusion of technologies that include high-speed internet, AI, advanced automation and robotics, big data analytics, and cloud technology. The reason why 4IR is so different from other industrial revolutions is because it is happening much faster and disrupting every industry in every country, and the breadth and depth of these changes 'herald the transformation of entire systems of production, management, and governance' (Schwab, 2016, p. 1).

According to Schwab (2015), the benefits of 4IR include raising global income levels and therefore improvements to quality of life for populations around the world. Other benefits include the transformation of the supply chain via nearshoring and onshoring through technological innovation with gains in efficiency and productivity, leading to a reduction in transportation and communication costs, as well as the opening up of new markets and economic growth (Schwab, 2015). It is also recognised that 4IR may result in inequalities through the disruption of the labour market by substituting human labour for automation and robotics, and this will give rise to a job market increasingly segregated into low-skill/low-pay and high-skill/high-pay segments, with a strong demand at the high and low ends but a hollowing out of the middle.

In terms of progress, the 4IR is quite unlike all preceding industrial revolutions as the technologies associated with this revolution are subject to three distinct waves of disruption forecasted for implementation within an incredibly short 16-year time frame from 2009 to 2025. Many of these technologies as summarised in Table 12.1 are extremely complicated and are still in development mode and beyond the capabilities of any individual such as 6G communication and quantum technologies.

According to Henrik Von Sheel, who is regarded as the originator of Industry 4.0 and the 4IR, there have already been two distinct waves of industrial disruption since its onset at the beginning of the millennium (TED, 2019).

The first wave was initially brought about by the 3IR involving digitalisation, advanced analytics, cloud computing, augmented reality, robotics, and 3D printing.

TABLE 12.1

The Three Waves of Technology Revolution

The Three Waves of the Fourth Industrial Revolution

1st Wave	2nd Wave	3rd Wave
2009–2016	**2016–2025**	**2025 +**
Digitalisation (internet of things)	Artificial Intelligence	Quantum Technology
Advanced Analytics	Autonomous Systems	Cybersecurity
Cloud Computing	Blockchain	Neurotechnology
Augmented Reality	Smart Automation	Nanotechnology
Robotics	6G Communications	Bioinformatics
3D Printing	Future of Energy	Advanced Materials

Source: Adapted from Von-Scheel (2022)

However, the use of many of these technologies has only recently begun to be adopted by the fashion industry accelerated by pressing global issues including the recent global pandemic (Kalypso, 2020). For example, the use of cloud computing using SaaS-based software has increased dramatically as the wider benefits of 3D prototyping software were realised during the pandemic when physical sampling was not possible.

The use of advanced analytics using AI have become an irreplaceable technology in support of sustainable fashion practice for many larger businesses involved in the fashion buying and retail sector in terms of its use for predicting and forecasting consumer demand for products and therefore the reduction of waste associated with traditional range planning and buying processes associated with 2IR and 3IR. It is forecasted that as software costs continue to reduce, AI will become more accessible to SMEs (Lu et al., 2022).

In terms of uptake of wave one robotics, with the exception of the sports shoe industry, uptake of robotics/cobots has only begun to infiltrate the sewn product industries and in particular apparel manufacturing in a superficial way via a variety of research and development projects as outlined in a separate chapter in this book. But this will change as according to McKinsey (2021) there are powerful reasons to develop UK manufacturing, including apparel manufacturing, which was valued at nearly £400 billion in 2019 and made up 9% of its GDP. Development of the sector will be dependent on advanced automation and robotics to accelerate affordable productivity, and many brands involved in manufacture have already begun to pilot these technologies, but most are slow to deploy at scale as the technology and the workforce are both still developing. Therefore, reskilling the workforce by retaining them is critical, and without concerted action McKinsey (2021) argues that two-thirds of the UK workforce will lack basic digital skills by 2030 while more than ten million people could be underskilled in leadership, communication, and decision-making.

Successful examples of upskilling are starting to emerge, for example Levi's has adopted the approach of upskilling existing employees to have high levels of digital literacy with ongoing training rather than training new digital recruits about fashion.

According to McDowell (2021), this helps to retain staff and preserve in-house functional knowledge whilst minimising the incidence of employees jumping ship or being enticed by unscrupulous recruitment strategies such as poaching by competitors.

The uptake of augmented reality, which was originally introduced to the fashion industry in 3IR for the purpose of augmenting apparel fitting, has experienced a renaissance of interest in 4IR both by the e-commerce sector for use in virtual reality scenarios such as Magic Mirrors but also by the developers of smart glasses to enrich the real-world experience of the user.

Finally, 3D printing within the sewn product industry has also proven successful within the footwear industry for the printing of running shoe soles and uppers; however its uptake and use for apparel still have major limitations as most TPU printing materials used in 3D apparel printing result in sculpture-style garments that cannot currently match the functional and comfort characteristics of traditional textiles used for most apparel designs.

All these new 4IR first-wave advancements require personnel to either develop these technologies or use them within particular functions of the fashion industry. Also, they need managers who can manage the users and the developers, and there is a shortage of all of the latter personnel. This is because uptake of these technologies has been slow and despite the fashion hype, many small-to-medium size enterprises have not yet put their foot on the rung of wave one technologies even though according to Von Scheel (2022) wave two commenced in 2016.

It is unquestionable that the fashion industry is behind in the adoption of many of these technologies. We have moved beyond 2023, some of the wave two and three technologies needed to drive 4IR such as 6G communications are just on the horizon. AI and Smart Automation are predominantly in pilot phase whilst blockchain is gaining momentum and is becoming increasingly important for tracking the fashion supply chain. For example, the UK Fashion and Textile Association recently launched a blockchain traceability project partnering with IBM and retailers including H&M, COS, Next, New Look, and several others to share information about the ethics and sustainability of their supply chains.

It is surprising that the uptake of some third wave technologies, such as nanotechnology and advanced materials, has become established prior to some wave two technologies whilst the success of other third wave technologies such as neurotechnology has been driven by concurrent developments and progression in the areas of AR Commerce using technology such as smart glasses that can be used to augment the process of virtual try-ons.

There are some limitations with smart glasses as they all still require touch-based access, but in the future, it may be possible to gain full control over smart glasses without touching any input devices using a brain–computer interface such as those in development with 4IR start-up NextMind, part of Snapchat, whose prototype headband lets its wearer gain control over several aspects of a computer. This includes, for example, aiming a weapon in video games or unlocking devices with just the power of one's mind. It seems almost unbelievable that you might be able to use your mind to make retail purchases in the near future.

Upskilling and recruiting for these new occupations are complicated for a number of reasons. First, as mentioned earlier, despite the hype the fashion industry is 12 years behind in adopting 4IR as many fashion businesses resisted digitalisation until forced to by the recent pandemic (Kalypso, 2020). Subsequently, the talent pool for 4IR is currently quite empty, and the race is on to source temporary skill via freelancers whilst upskilling existing workforces. The higher education sector is also behind in integrating 4IR technologies into their curriculums. Many fashion schools in the UK have developed links and partnerships with the same fashion businesses who resisted digitalisation, and many are therefore still delivering fashion courses that meet the employment requirements for IR2 and early IR3. More positively, some of the top fashion schools, such as Manchester Metropolitan University and London College of Fashion, have made good inroads both during and after the recent pandemic in adapting the curriculum for late 3IR and 4IR phase one.

To clarify how digitalisation is disrupting careers in the fashion industry, the following section will outline some of the new and emerging 4IR fashion industry jobs, including the knowledge and skills base they require whilst finally outlining what can be done to support the development of the fashion industry's workforce of the future.

12.6   WORKFORCE OF THE FUTURE

The 4IR is without doubt changing the way we live, work, and how we relate to one another. It is different from earlier industrial revolutions because although uptake of these technologies has been relatively slow within the fashion industry, it is happening much faster than any previous industrial revolution. The fashion industry is utilising digitalisation and advances in sewn product technology to transform the entire system of production, management, and governance. This immense change is forecasted to result in inequalities through the disruption of the labour market by substituting human labour for digitalisation involving automation, AI, and robotics. This will give rise to a job market increasingly segregated into low-skill/low-pay and high-skill/high-pay segments, with a strong demand at the high and low ends but a hollowing out of the middle (Schwab, 2016).

According to McKinsey (2018), demand for higher cognitive skills, such as advanced IT skills and programming, advanced literacy skills, creativity, critical thinking, problem-solving, and complex information processing, will grow through 2030. The next wave of automation and AI will disrupt manufacturing jobs through better analytics and increased human–machine collaboration. The number of professional jobs such as sales representatives, engineers, managers, and executives are expected to grow. The need for technological skills, both advanced IT skills and basic digital skills, will increase as more digital technology professionals are required, this is also true for the fashion retail sector where physical and manual skills along with basic data processing will decline, replaced by a surge in demand for advanced IT skills involving complex information processing skills (McKinsey, 2018).

To fill jobs that are created at the high skill/high pay areas, employers will need to adopt two approaches. First, many companies will employ more temporary freelancers to fill immediate gaps in their workforce. Second, this latter approach will enable businesses to develop their existing staff to transition into some of these gaps, and as McKinsey (2019) argues, the staffing gaps won't just be in specialist areas such as digital technology or engineering, it will also involve shortages of management needed to lead change across all jobs in the supply chain. In statistical terms, it is estimated that up to ten million workers could be underskilled in the areas of leadership and management, and 30% of all UK workers may need to transition between occupations or skill levels by 2030 (McKinsey, 2019).

The level of disruption that is and will be caused by these changes will have a profound impact on job roles in the fashion industry, some of which may seem unimaginable right now such as user experience designer, holographic specialist, or digital knowledge manager whilst others are already beginning to appear on the horizon such as 3D fashion designer or avatar/human body specialist. This point is supported by the FT Alliance (2021), which argues that the ongoing digitalisation of the fashion industry is creating new business models and therefore new jobs requiring new knowledge and new skillsets most of which are at the high-skill/high-pay end of the job market. They have identified eight new *job families* encompassing up to 50 new fashion-related jobs. Their detailed findings are summarised in Table 12.2.

The transformation of the fashion industry will refocus it as an industry that is dependent on higher-level qualifications, lifelong learning, and CPD. According to McKinsey (2018), individuals with a college degree are more likely to be hired or contracted, more likely to receive retraining, and less likely to be displaced. However, demonstrating the right mindset and the right mix of soft skills such as being entrepreneurial, open to change, a team player, flexible, and adaptable will potentially become more important than simply demonstrating a *specific digital skills and knowledge set* (FT Alliance, 2021, p. 53).

The responsibility for building the workforce of the future will sit with a variety of stakeholders, but the two key players include fashion businesses and the higher education sector whose key role between now and 2030 will be to successfully transition ten million people from manual or administrative roles which will be replaced by AI and automation to leadership and management and technical roles requiring higher cognitive skills (McKinsey, 2019).

A starting point in facilitating this transition will be to proactively upskill the existing workforce by identifying employees' hidden talents rather than solely considering qualifications and employment history. This approach is being spearheaded by Levi's who is focusing on developing the digital skills of its existing workforce using data science bootcamp approach rather than employing new staff where there may currently be talent shortages and teaching them about fashion which it has reported enhances employee satisfaction and helps to hold onto reliable and valuable staff who otherwise might look for new opportunities to develop in competitor businesses (McDowell, 2021).

A summary of possibilities for upskilling and training individuals or employees with new skills include:

- Industry partner with chosen HEI to develop bespoke educational programmes
- Short courses delivered by HEIs
- Internal training using peer-to-peer learning
- Buying short courses from specialist companies
- Open source online virtual learning using, for example, YouTube videos
- Online virtual learning using companies such as LinkedIn learning or Udemy
- University degree courses

TABLE 12.2

Future Job Roles in the Fashion Industry

Name of Job Family	Description of Job Family	Examples of Job Roles in This Family
Innovation Manager	A leader who has a clear understanding of the digitalisation of the fashion industry and can manage the strategy for business transformation	Digital Product Manager, Creative Technologist, Digital Knowledge Manager, Transformation Specialist, Engineer, Designer, Educator
UX–User Experience Designer	The ability to create successful and sustainable on-demand products using a mix of traditional and digital skills for specific end users/markets	3D Modelling Specialist, 3D and visual expert, Virtual 3D Imaging (updated from photographer), Avatar/Human Body Specialist, AR/VR Expert, Holographic Specialist, Digital Product Tester, 3D Pattern Maker, Pattern Programmer/Designer, Zero Waste Pattern Maker/Designer, Industrial Designer (3D Modeller)
Omnichannel and E-Commerce Skills and Roles	Concerned with developing and evaluating consumers' digital experience and their needs and how best to meet them	Expert in 3D E-Commerce, User Experience Designer, Digital Experience Manager, Customer Experience Designer, Customer Success Manager, Vendor Integration Specialist, Creation Platform Manager, Personal Tailor, Virtual Seller
Sustainability Lead	Concerned with developing and delivering sustainability strategies and to influence and maintain change by working with key stakeholders to set ambitious targets	Circular Design Pattern Expert, Chemical Designer (Scientist), Chemical Engineer, Green Fabric Sourcer, Fabric Component Designer, Material Researcher, Material Innovation Manager, Ecosystem Innovation Manager, Eco Fabric Designer
Digital Product Manager	Involved with designing, managing, creating, and maintaining product documentation whilst collaborating with key stakeholders to identify, define, and design solutions to improve the consumer experience	Systems Designer, Innovation Manager (working with product development teams)
Data Analysis, Management, and governance	This area centers around the role of data scientist dealing with large data sets; background qualifications are in the area of Computer Science, physics, maths, data analytics	Data Scientist, Data Analyst, AI Expert, Software Developer
Lobbyist	This area involves persuading legislators to vote on public policy in favour of their clients' interests, requiring strong communication and analytical skills	Policy Influencer
Micro-Factory Manager	Working with cutting-edge technologies with knowledge and experience across a broad range of manufacturing operations functions	Head of Technology, Factory Manager, 3D Printing Technologist, Advanced Automation Engineer

Source: Adapted from the FT Alliance (2021)

HEIs will continue to play a major role in ensuring jobs of the future will have a steady stream of new graduates, who can progress into professional employment or further study, with qualifications that hold their value over time (OFS, 2022).

There are a number of key areas where HEIs involved in the delivery of fashion education need to refocus attention. First, according to McKinsey (2018), they need to build more partnerships with fashion businesses and vice versa as their research has concluded that industry currently favour internal training over HEI training.

HEIs will also need to work closely with fashion businesses to capitalise on student placements and internships as this enables a two-way sharing of new skills and knowledge. According to Mitchell (2022), students with relevant 4IR skills can be matched with companies for work placement to help the organisation understand the tangible benefits of digital technologies quicker without having to make permanent employment commitments. Second, HEIs also need to ensure that the curriculum is flexible and relevant so that graduates exit with a future-proofed skill set underpinned by a flexible lifelong learning mentality. This becomes more relevant when we consider that 4IR involves three waves of new technologies and presently, most fashion businesses are still behind or getting to grips with wave one technologies before wave two and three begin to wash over them from 2025 onwards. Third, HE will need to focus on developing their CPD initiatives so that appropriate time for continuous upskilling of their own staff becomes a strategic priority (Mitchell, 2022).

REFERENCES

Dean, P (2000) *The First Industrial Revolution*, Cambridge University Press, Cambridge

Ferrell, O, et al. (2015) *Understanding the History of Marketing Education to Improve Classroom Instruction* [online], www.researchgate.net/publication/281045375_ Understanding_the_History_of_Marketing_Education_to_Improve_Classroom_ Instruction (Accessed 03/01/2024)

FT Alliance (2021) *Fashion-Tech Jobs Profiles Portfolio: Future Job Roles in Fashion-Tech and Human Resource Guidelines and Future Recruitment Tools* [online], www.arts. ac.uk/knowledge-exchange/stories/new-report-unveils-8-future-jobs-in-fashion-tech (Accessed 03/01/2024)

Hayes, SG and McLoughlin, J (2008) Technological Advances in Sewing Garments, Chapter 10. In: Fairhurst, C (ed.) *Advances in Apparel Production*, Woodhead, Cambridge

Honeyman, K (2000) *Well Suited: A History of the Leeds Clothing Industry*, Pasold Research Fund, Oxford

Howes, A (2017) *The Relevance of Skills to Innovation During the British Industrial Revolution, 1547–1851* [online], www.antonhowes.com/uploads/2/1/0/8/21082490/ howes_innovator_skills_working_paper_may_2017.pdf (Accessed 03/01/2024)

Kalypso (2020) *The 2020 Digital Product Creation Survey Briefing* [online], https://kalypso. com/files/docs/Exec-Summary-Annual-Retail-Innovation-Adoption-Survey-2020_ 2020-12-17-215126.pdf (Accessed 27/12/2023)

Lasi, H, Fettke, P, Kemper, HG, et al. (2014) Industry 4.0. *Business Information Systems Engineering Journal*, 6(April): 239–242 [online], www.researchgate.net/publication/ 271950998_Industry_40 (Accessed 03/01/2024)

Lu, X, et al. (2022) *Ai-Enabled Opportunities and Transformation Challenges for SMEs in the Post-pandemic Era: A Review and Research Agenda. Front Public Health* [online], www.ncbi.nlm.nih.gov/pmc/articles/PMC9098932/ (Accessed 03/01/2024)

The Manufacturer (2019) *Henrik von Scheel: In Conversation with the Father of Industry 4.0*, [online], www.themanufacturer.com/articles/henrik-von-scheel-in-conversation-with-the-father-of-industry-4-0/#:~:text=Alumni%202021%2F22-,Henrik%20von%20Scheel%3A%20In%20conversation,the%20%27Father%20of%20Industry%204.0%27&text=Success%20may%20have%20many%20parents,a%20long%2Dterm%20competitive%20advantage (Accessed 03/01/2024)

McDowell, M (2021) *Lessons from Levi's Data Science Bootcamp, Vogue Business* [online], www.voguebusiness.com/technology/exclusive-lessons-from-levis-data-science-bootcamp (Accessed 03/01/2024)

McKinsey (2018) *Skill Shift: Automation and the Future of the Workforce* [online], www.mckinsey.com/featured-insights/future-of-work/skill-shift-automation-and-the-future-of-the-workforce (Accessed 03/01/2024)

McKinsey (2019) *The Future of Work: Rethinking Skills to Tackle the UK's Looming Talent Shortage* [online] https://www.mckinsey.com/featured-insights/future-of-work/the-future-of-work-rethinking-skills-to-tackle-the-uks-looming-talent-shortage (Accessed 20/02/2024)

McKinsey (2021) *Facing the Future: Britain's New Industrial Revolution* [online], www.mckinsey.com/capabilities/operations/our-insights/facing-the-future-britains-new-industrial-revolution (Accessed 03/01/2024)

McKinsey (2022) *How Technology Is Shaping Learning in Higher Education* [online], www.mckinsey.com/industries/education/our-insights/how-technology-is-shaping-learning-in-higher-education (Accessed 03/01/2024)

Michon, H (2022) *Industrial Revolution, the Economic Historian* [online], https://economic-historian.com/2021/03/industrial-revolution/ (Accessed 03/01/2024)

Mitchell, A (2022) cited in Holden, L (2022) *How to Create a Future Proof Fashion Team for Digital Transformation* [online], https://fashionunited.com/news/business/how-to-create-a-future-proof-fashion-team-for-digital-transformation/2022073148920 (Accessed 03/01/2024)

OFS (2022) *Maintaining the Credibility of Degrees* [online], www.officeforstudents.org.uk/publications/maintaining-the-credibility-of-degrees/ (Accessed 03/01/2024)

Roberts-Islam, B (2020) *Is Digitization the Saviour of the Fashion Industry* [online], www.forbes.com/sites/brookerobertsislam/2020/01/07/is-digitisation-the-saviour-of-the-fashion-industry-i-ask-a-cto-who-knows/?sh=749287f87e7a (Accessed 03/01/2024)

Schwab, K (2015) *The Fourth Industrial Revolution: What It Means and How to Respond* [online], www.weforum.org/agenda/2016/01/the-fourth-industrial-revolution-what-it-means-and-how-to-respond/ (Accessed 03/01/2024)

Schwab, K (2016) *World Economic Forum, the Fourth Industrial Revolution* [online], www.weforum.org/about/the-fourth-industrial-revolution-by-klaus-schwab/ (Accessed 03/01/2024)

TED (2019) *Creating the Innovation Spark* [online], www.ted.com/playlists/11/the_creative_spark (Accessed 03/01/2024)

TEM (2022) *Technical Education Matters, Developments between 1920 and 1940* [online], https://technicaleducationmatters.org/2009/06/28/chapter-10-developments-between-1920-and-1940/ (Accessed 03/01/2024)

Temperton, J (2015) *A Fourth Industrial Revolution Is about to Begin (In Germany)* [online], www.wired.co.uk/article/factory-of-the-future (Accessed 03/01/2024)

Tyler, D (2009) *Carr and Lathams Technology of Clothing Manufacture*, Blackwell, London

Von Scheel, H (2022) *Three Waves of Technology Revolution* [online] https://www.researchgate.net/publication/333450638_Demystify_the_Industry_40_and_move_beyond_hype (Accessed 20/02/2024)

Windrush Team (2022) *Windrush Stories* [online], www.windrushday.org.uk/news/howcaribbean-migrants-helped-to-rebuild-britain/ (Accessed 03/01/2024)

13 Vorteq Case Study
High-Performance Cyclewear Product Development

13.1 INTRODUCTION

This case study will present complex considerations in the design process employed by TotalSim company Vorteq in developing a custom-built overshoe for elite cyclists, with a focus on design and technology to improve fit, enhance aerodynamics, reduced resistance, and increased speed. For the purpose of this case study, the unique challenges in design to optimise aerodynamic flow will be examined through the lens of competition cycling and the design and development process for a custom-fit overshoe. The chapter will commence by outlining the challenge of creating performance-enhancing sportswear as well as the challenges of designing apparel in the highly specialised area of aerodynamics. The chapter will move on to consider the design and development process utilised for the development of this product including the unique six-step process that was followed during the development of the Tokyo Edition, including scanning, draping, design exploration, and exploitation involving compliance and material, wind tunnel testing, and final product completion.

The key contributors to this chapter include Dr Rob Lewis (OBE), Founder of Vorteq and Managing Director of TotalSim, leaders in aerodynamics and computational fluid dynamics, and Jane Ledbury, an emeritus apparel industry expert and Principal Lecturer in Fashion Design and Product Development at Manchester Metropolitan University.

13.2 THE CHALLENGE OF CREATING PERFORMANCE-ENHANCING PERFORMANCE SPORTSWEAR

The chances of winning a gold medal are 22 million to one.

(Brownlie, 2011)

According to Vorteq founder Dr Rob Lewis the challenges around cycling aerodynamics are harder than Formula 1 cars (TotalSim, 2017). This case study will examine the design and development process employed by Vorteq in solving complex problems to increase marginal gain for elite athletes. The combination of *fast* materials and customisation delivers an overall drag reduction of between 2% and 4% for the athlete. When combined with a custom-made Vorteq skinsuit, aerodynamic gain can be even greater and as much as 9%.

Brownlie et al. (2016) suggests that elite athletes have similar focus, practice regimes, and support systems, and therefore performance variations and margins

DOI: 10.1201/9781003126454-13

at this level are minimal, with the differences between achieving either gold or silver medals often being a fraction of a second. Consequently, advantages are sought through advances in technology, apparel, and equipment. Although elite athletes may have similar training, focus, and support teams, each athlete has a unique and distinctive physiology, and according to Lewis (2021), to gain maximum advantage for the wearer each apparel product is bespoke and engineered specifically to suit the athlete's body shape, predominant posture, sports discipline, event, and prevailing conditions.

At this level of competition clothing and equipment can provide the crucial difference between the first and second place. This was demonstrated by Speedo's development of the LZR Fastskin swimsuit, which was considered by critics to have provided unfair advantage to users at the Beijing Olympics in 2008. Athletes wearing the controversial swimsuit broke 74 world records between its launch in March and November 2008, resulting in the term 'technological doping' (Kessel, 2008) in reference to clothing and governing body FINA subsequently banning high-tech 'super suits' from 2010.

The swimsuit was developed by the company's Aqualab, with research and development, which included a team of designers, textile engineers, sports scientists, biomechanical engineers, physiologists, physiotherapists, and over 100 elite athletes, and was tested in a wind tunnel at NASA. The swimsuit crucially reduced drag and shaped the body through compression to become more aquadynamic, allowing the swimmer to move through the water more easily and therefore with greater speed (Kessel, 2008).

Brownlie et al. (2016) claims that 'the only way to go faster is to reduce resistance', this case study will examine how the British company Vorteq engineers apparel to enhance aerodynamics, reduce resistance, and cut drag to gain competitive advantage for the wearer.

13.3   ABOUT VORTEQ

Vorteq is a TotalSim company based within the TotalSim Silverstone Technology Cluster, housing a team of experts in computational fluid dynamics (CFD). Their services include consultancy, support, and training in CFD across a broad range of applications worldwide. Vorteq leads development in performance sport aerodynamics and enables TotalSim's elite equipment to be made available to the wider consumer market (Vorteq, 2021).

Dr Rob Lewis OBE, Managing Director of TotalSim Ltd., which he founded in 2007, holds a PhD in CFD and is Fellow of the Institute of Mechanical Engineering and the Institute of Engineering and Technology. Dr Lewis led in CFD use in Formula One in the 1990s and subsequently with British American Racing and Honda F1. In 2017, he was appointed an Officer of the Order of the British Empire, for services to science applied to sport in recognition of his contribution to the performance of British athlete's performance over the last three Olympic campaigns (TotalSim, 2017).

TotalSim has worked with British Cycling, Team Sky, Aston Martin, F1 teams, and the English Institute of Sport (TotalSim, 2021). Dr Lewis founded Vorteq in 2019 with the aim 'to create the best possible sporting equipment on the planet, with no compromise to performance' (Vorteq, 2021). Vorteq engineers have worked with elite athletes to develop sports performance technology for more than a decade and provide cutting-edge bespoke apparel products aimed to optimise marginal gain (Vorteq, 2021). The company's specialism and expertise lies in the area of aerodynamics and the reduction of drag for incremental gain needed to succeed at competition. Vorteq works with athletes across a number of disciplines such as track and road cycling, skiing, and bob sleigh where aerodynamic efficiency is crucial. The company developed skinsuits and overshoes for a number of Olympians achieving medals in the Tokyo 2021 Olympics, providing skinsuits for national teams, including those of Japan, Malaysia, and Trinidad and Tobago, and overshoes for many more athletes.

13.4 DESIGN CHALLENGES FOR PERFORMANCE: AERODYNAMICS

Vorteq's expertise lies in a quantitative understanding of aerodynamics and the challenges it presents to human performance where maximum speed is critical. Vorteq provides Human Performance Services, which enable athletes to establish optimal biomechanical position and their interaction with airflow to reduce drag that inhibits speed.

When a solid object travels through air it interacts with millions of air molecules, which create complex forces (Airshaper, 2022). These forces can be experienced when riding a bicycle, even at normal or slow speeds, with half of the energy and effort expended by the rider used in overcoming aerodynamic drag and the greater the velocity, the greater the increase in drag. It is estimated that approximately 90% of mechanical energy is required to overcome aerodynamic drag (Debraux et al., 2011).

Drag experienced when travelling through air falls into two distinct categories, pressure drag and friction drag. Minimising the effect of air resistance and drag, particularly for competition purposes, can gain important advantages in both speed and performance (Lewis, 2021; Airshaper, 2022).

Pressure drag is the force impacting the frontal area of the athlete, which builds both force and pressure; additionally, as air travels along an athlete's surface a wake is created, increasing pressure drag and slowing down the speed of travel. Areas contributing to pressure drag are reduced by the rider's posture on the bike at speed generally in an aero position, with the greatest area of frontal impact to the helmet, hands, and legs.

Friction drag is created by air passing tangentially over the surface of the athlete as they move through airflow. As air slides over a surface, it creates friction forces, known as friction drag and develops a boundary layer at the athlete's apparel surface (D'Auteuil et al., 2010). Generally, pressure drag makes a greater contribution than friction drag to aerodynamic resistance, front-facing surfaces generate no frictional drag as air impacting these surfaces decelerates to zero (Airshaper, 2022).

Airflow creates a laminar boundary layer when passing over smooth surfaces creating an area of slow-moving air, which becomes denser and needs to contract and separate from the surface; once separation occurs a wake is created, increasing pressure drag and reducing velocity.

The use of surface textures or texture morphology creates a turbulent boundary layer, which interacts with air surrounding the cyclist and maintains kinetic energy.

The kinetic energy generated by the turbulent boundary layer is retained enabling airflow to attach to the surface for a greater duration, thereby reducing wake and consequently pressure drag (Airshaper, 2022). This principle is applied to textiles and apparel in the strategic placement of surface textures to reduce aerodynamic drag and enhance performance for the user.

Wind tunnel trials and CFD analysis used in the development of Vorteq sports apparel can establish the complicated laminar and turbulent airflows created by the athlete's unique body morphology and riding position to identify airflow separation and inform strategic placement of texture to apparel design and development.

Amongst considerations in reduction of drag is the cyclist's coefficient of drag area (CDA), which is quantified in square metres. CDA is used to compile factors of airflow, shape, and position relative to the cyclist and is the quotient between dynamic pressure of airstream and the presented frontal area and aerodynamic drag (D'Auteuil et al., 2010). Importantly, when measuring CDA, the lower the measurement, the more aerodynamically efficient the cyclist; however; extreme aerodynamic cycling position may impact upon power, and wind tunnel trials enable the athlete to establish optimum position for both speed and duration.

13.5   DESIGN AND DEVELOPMENT PROCESS

The Vorteq research and development method is a six-step process consisting of scan, drape, design, assembly, wind tunnel testing, and final product. Design and development of each custom product is based upon expertise in textile performance, construction of materials, surface texture, and aerodynamic performance.

The process begins with a full body scan capturing the athlete's predominant position on their bike. The data captured is processed and provides an accurate virtual avatar of the athlete, which is then 3D digitally printed to full scale for physical testing in the wind tunnel. The second step in the research and development process is drape, where the 3D virtual model is used to create patterns, which will provide perfect fit to the athlete's unique physiology for optimum performance.

Design constitutes the third step, with options, which include team colours and logo, an athlete's individual design requirement or designs developed by Vorteq. Design considerations are made in respect of fabric texture and pattern. Assembly, or step four of the research and development process, is carried out in the Vorteq Materials Lab, based at TotalSim in the Silverstone Sports Engineering Hub. Each fabric section is sublimation printed and the patterns cut prior to hand assembly by garment technicians in the Lab.

The fifth step in the process is wind tunnel testing. In the early stages of the development process, a number of variations in materials and product are tested in a small wind tunnel in the Materials Lab to establish performance levels and efficiency. Once established, the athlete's event, position, wind speed, and angle are closely replicated, and products are tested in the main wind tunnel on the full-scale model to gather data on product performance and determine drag reduction/marginal gain.

The sixth and last step in the process is the final product. This is produced following an initial prototype and is the result of a number of tests to establish optimum fabric, texture, and design combinations for maximum performance gain for the athlete (Vorteq, 2021).

Each of these processes will be examined in greater detail with particular reference to the technologies exploited to produce bespoke high-performance sports products to optimise marginal gain.

13.6   RESEARCH AND DESIGN

Vorteq presented the 'Ultimate Performance Skinsuit' for the Tokyo 21 Olympics, and the custom skinsuits were worn by medal-winning athletes. According to the company, the custom-built skinsuit was the result of a £400,000 investment in research and development, with over 40,000 combinations of materials and fabric subjected to wind tunnel testing resulting in the 'Tokyo Edition' skinsuit. Aimed at maximising track cycling speeds, each custom-built skinsuit and overshoe is the outcome of exhaustive wind tunnel testing in consideration of materials and textures as detailed earlier and the athlete's unique body morphology and position on the bike. Lewis (2021) describes the continuous development process as 'valley hopping', following incremental experimentation, it is necessary to 'hop out of the valley you are in' and make radical design changes, looking at new ways of thinking and new approaches to make significant gain to further reduce drag.

13.6.1   Vorteq Six-Step Process for the Development of the Tokyo Edition

Step 1—Scan: In creating the Tokyo Edition, Vorteq used an Artec Leo handheld scanner to capture the athlete's body and position in high resolution 3D. The Artec Leo scanner was used to produce 3D digital life-sized models to be used in wind tunnel testing, thereby reducing the time an athlete must be physically present at the TotalSim facility. Following initial tests on the life-sized 3D mannequin, ultimate testing is carried out with the athlete in the main wind tunnel. Data gathered from the handheld Leo scanner is combined with CFD analysis, CAD skills, and design thinking to produce the optimal sports product (Vorteq, 2021). CFD can be used in simulating airflow to establish where air pressure is impacting on a surface and in finding solutions to issues of airflow that affect performance.

Competition cycle clothing must fit closely to aid aerodynamic efficiency and the Vorteq skinsuit aims to reduce CDA (drag coefficient × area) significantly in comparison to general cycle apparel (McMillion, 2020). The handheld Artec Leo, a portable cable-free scanner, is designed for accuracy, manoeuvrability, and speed, capturing medium-sized objects in high resolution and enables Vorteq to gain 3D images of an athlete in situ on their bike in predominant posture or racing position. TotalSim, the parent company of Vorteq, was the first in the UK to use the Artec Leo scanner in 2019, and Vorteq was quick to recognise its potential to increase speed and efficiency in design workflow for their skinsuit (McMillion, 2020). The award-winning wireless Artec Leo scanner provides onboard automatic processing, with the touch panel screen displaying the 3D object being built enabling the user to check complete capture whilst simply walking around the object and holding the scanner in one hand (Artec, 2021). The Artec Leo scanner enables highly accurate scans, with 3D point accuracy, up to 0.1mm and capture times of a rider in the cycle position of just 5–6 minutes.

FIGURE 13.1 3D scanning athlete (2021).

Source: Courtesy of Vorteq, www.Vorteqsports.co.uk

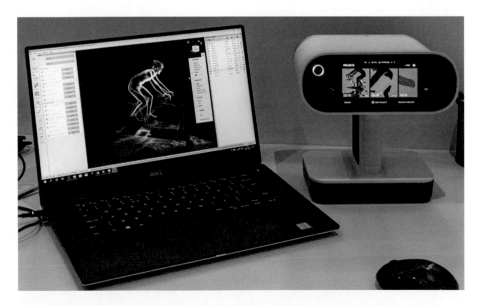

FIGURE 13.2 Artec Leo 3D scanner with Artec Studio software displaying scan of rider at Silverstone Sports Engineering Hub.

Source: Courtesy of Artec (2021), www.Artec3D.com

The Artec Leo scanner employs two powerful 3D programs and AI, using the very latest 3D algorithms. Its touchscreen interface is used for scanning and real-time processing. When scanning is completed, the data can be transferred to 3D software Artec Studio for editing, processing, and analysis (Artec3D, 2021). Vorteq will then carry out a Smooth Fusion, the algorithm used to process scans that have undergone movement such as body scans for skinsuits and overshoes as data is smoothed during the running of the algorithm, producing clean and accurate models. For inanimate objects, such as a bike, a Sharp Fusion is used to capture geometrical data (LaserDesign, 2021).

3D digital printing, or additive manufacturing, is a process which creates a physical object from a digital file by laying down multiple layers of material (Ultimaker, 2022). It is particularly valuable in customisation and has the added advantages of speeding up development processes, known as rapid prototyping and eliminating waste. To process the scans with speed, Vorteq uses a bank of 12 large 3D digital printers to print multiple body pieces, subsequently joined together with both magnets and bolts to provide an exact replica of the athlete in predominant posture on their bike. Using life-sized 3D mannequins provides an opportunity to conduct multiple tests without requiring the athlete to be present. Additionally, in producing extra parts a package can be provided to include multiple leg and hand positions for testing (Vorteq, 2021). Vorteq uses life-sized 3D models of the athlete and CFD analysis, together with wind tunnel testing to establish the kinesiological riding practices of the athlete to collate data to create the optimal apparel for the cyclist (McMillion, 2020).

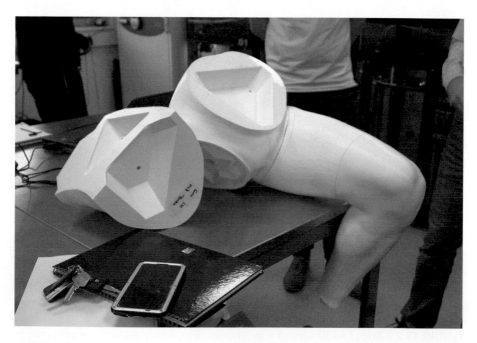

FIGURE 13.3   3D digital printed athlete's body (2021).

Source: Courtesy of Vorteq, www.Vorteqsports.co.uk

Step 2—Drape: Using a 3D model of the athlete, designs are developed by draping on an avatar of the athlete. The virtual garment can be evaluated and adjusted at this stage to achieve optimum fit, taking into consideration the particular properties of the fabric, such as degree of elasticity. 2D patterns are created through a process of 'flattening' 3D to 2D pattern pieces, and seams etc. are then added, and the 2D pattern is sent to the printer in preparation for cutting fabric for the prototype. This process eliminates the need to conduct some physical fit tests, thereby reducing both development time and material waste, thus contributing to sustainable practice. As athletes have unique physiologies, this method enables Vorteq to efficiently optimise fit for performance (Lewis, 2021).

Step 3—Design: Exploration versus exploitation—Vorteq, having scanned the athlete, printed a life-sized 3D model of the athlete, and draped bespoke patterns, proceeds with *Design of Experiments*, where a variety of technical fabrics are tested on a cylinder resembling the lower leg in a small wind tunnel to assess performance with reference to a number of parameters such as length, width, seam placement, seam height, zip positioning, and optimum trip position. Results are analysed to establish aerodynamic efficiency, prior to proceeding to sample assembly and testing on the 3D model and ultimately the athlete in the main TotalSim wind tunnel.

13.6.2 COMPLIANCE

All design and development decisions are made in reference to compliance with the rules and regulations set out for competition by Union Cycliste Internationale (UCI), the world governing body for sports cycling. The UCI oversees all international competitive cycling events and has stringent rules pertaining to cycle clothing. Infringement of the rules for a particular event can result in a ban from competition or fines for either the individual or the team in question.

The UCI's rules state that

> Items of clothing may not modify the morphology of the rider and any non-essential element or device, of which the purpose is not exclusively that of clothing or protection, is forbidden. This shall also apply regarding any material or substance applied onto the skin or clothing and which is not itself an item of clothing.
>
> (UCI, 2021)

An example which demonstrates the importance of UCI rules involves the Danish Olympic team who were at the centre of a controversy at the 2021 Tokyo Olympics, when all members of the team presented with kinesiology tape in identical positions on both shins, placed below the sock and extending to just below the knee. As this was unlikely to be due to matching injuries, it was intimated that the tape was applied to the front of the lower leg and could trip the boundary layer in creating turbulence and therefore enhance performance. As a result, the UCI banned the tape and issued the Danish team with a warning (Croxton, 2021).

Regulations pertaining to apparel, including the overshoes, may vary according to the sporting event and can be exacting; for example, rules for track cycling specify that socks and overshoes used in competition may not rise above the height defined by half the distance between the middle of the lateral malleolus and the middle of the fibula head (Ballinger, 2018). Essentially, an overshoe or sock must not be higher than the midway between the ankle and base of the knee. Regulations for road cycling, time trial events, and triathlon may vary.

13.6.3   Materials

The Vorteq overshoe needs to fit closely for aerodynamic efficiency, therefore fabrics include fibres such as Lycra to provide the degree of stretch required to achieve optimum fit. Experimentation includes both knit and woven materials in strategically placed zones to establish maximum performance. The primary purpose of the overshoe is to reduce drag and the energy expended through air resistance towards aerodynamic gain; however, technical textiles can offer a number of additional benefits, including breathability and prevention of water ingress, allowing the rider to remain warm and dry. As described previously, texture or shape of fabric surface on both the skinsuit and overshoe is crucial as this *trips* the boundary layer reducing drag. Design of experiments typically includes testing a number and variety of weaves, textures, and their strategic positioning to establish the best possible marginal gain for each athlete (Lewis, 2021).

13.6.4   Design Features

The Vorteq overshoe requires particular attention to fabric, fabric texture and fabric tension, zip positioning and placement, height, and thickness of the trip. The *trip*, or *trip strip*, is designed to enable the calf to slip through air more efficiently. This interruption to a smooth surface artificially 'trips the boundary layer', by converting the air meeting the lower leg from laminar to turbulent flow, enabling the air to attach to the leg for longer time before separation and consequently reducing drag. According to AeroCoach (2021), correctly positioned trip strips can save an athlete as much as five watts or several seconds over a ten-mile time trial. Overshoes may also vary according to event, and specific challenges such as speed and wind angle are taken into consideration. Lewis (2021) suggests that with wind angle from the right, a 'right hand' pair of overshoes may be required to counter sail effect and create thrust. Additionally, speed variation for pursuit with a rate of 15–16 metres per second may require a different intervention to sprint events with a speed of 19–20 metres per second.

Donning and doffing are a particular consideration as is common with the skinsuit the garment must fit closely and stretch fabrics are typically used to optimise fit. Consequently, putting the overshoe on and taking it off must be carried out with care to avoid damage to the product.

Step 4—Assembly: Assembly of test pieces, samples, prototypes, and final product is conducted at the Vorteq Materials Lab based at TotalSim in the Silverstone Sports Engineering Hub. Each fabric section is printed prior to hand assembly by garment technicians in the Lab. Construction includes both sewn and non-sew technology in the form of adhesive bonding, reducing both bulk and water ingress.

Step 5—Wind Tunnel Testing: Design of experiments-testing typically begins with interventions on a variety of materials on a cylinder in the small wind tunnel to establish their aerodynamic properties and then progresses to trials of different fabric combinations and tensions, seams, zip and trip placement for foundations and trip optimisation. Testing on cylinders, which combines the factors described in many permutations, is a proven methodology to establish best amalgamation of component parts and textiles as confirmed by Hong and Asai (2021), who found that when testing on cylinders, that represent an arm, the aerodynamic forces of various textiles results demonstrated that air resistance altered with fabric surface morphology and that a reduction of air resistance of up to 8% was possible. Further, the authors proposed that bespoke suits in consideration of event conditions could additionally improve performance.

FIGURE 13.4   The Silverstone Sports Engineering Hub Sports Performance Wind Tunnel (2021).

Source: Courtesy of Vorteq, www.Vorteq.co.uk

FIGURE 13.5   3D scanning of a rider in the Sports Performance Wind Tunnel at SSEH.

Source: Courtesy of Artec (2021), www.Artec3D.com

FIGURE 13.6 A rider in the wind tunnel at the Silverstone Sports Engineering Hub (2021).

Source: Courtesy of Vorteq, www.Vorteq.co.uk

FIGURE 13.7 The final product Vorteq Overshoe.

Source: Author's own (2022), courtesy of Vorteq www.Vorteq.co.uk

Subsequently, testing progresses to developed product on a 3D model of the athlete in the main wind tunnel, which may present several different challenges. Although 3D digital modelling of an athlete in full scale and in predominant posture provides great opportunity to repeat multiple tests, without requiring an athlete to be physically present, complexities include testing a skinsuit or overshoe on a 3D model, which is static and not representative of movement and pedalling activity.

Finally, testing is conducted in the wind tunnel on the athlete in motion, which presents challenges for repeatability of test results in terms of *noise*, or the differentiation of results can be challenging to establish the optimum; for example, if a cyclist in the wind tunnel might measure a drag of 40 newtons, dismounts and remounts, and this changes the drag to 40.5 newtons, then later puts on the same suit when the drag become 41 newtons. Consequently, repeatability of experiments can vary by up to 2%. Separating out each of the parameters can prove difficult, and Vorteq technicians must strive to see through the *noise* or variation of test results due to these uncertainties (Lewis, 2021).

Step 6—Final Product: Each final product is bespoke to the athlete and is the result of extensive development following the design of experiments to establish maximum gain and optimum positioning for the sport, event, and conditions. The skinsuit or overshoe has been specifically engineered and custom-built in the fastest materials for the athlete having established their optimum dynamic profile.

FIGURE 13.8 Founder of Vorteq and Managing Director of TotalSim, leaders in aerodynamics and computational fluid dynamics, in front of the Wind Tunnel Silverstone UK (2021).

Source: Author's own

13.7 SUMMARY

It is evident that the significant investment made by Vorteq in terms of both finance and development has succeeded in placing the company at the forefront of cutting-edge apparel technology aimed at maximising aerodynamic gain, realising the brand's ambition to 'to create the best possible sporting equipment on the planet, with no compromise to performance' (Vorteq, 2023). Future development will consider wind angle and prevailing conditions for greater marginal gain increasing competitive edge for the athlete.

As leaders in the field of design for aerodynamic efficiency and the wealth of knowledge and expertise in the area of CFD, together with the first-class facilities at the TotalSim Silverstone Technology Cluster (Silverstone, 2021), Vorteq is set to remain at the forefront of progressive research and development in support of athletes and their aspirations in achieving the ultimate gold medal.

REFERENCES

AeroCoach (2021) *Learn About Aerodynamics* [online], www.aero-coach.co.uk/learn (Accessed 02/01/2024)

Airshaper (2022) *What Is Aerodynamic Drag, Pressure Drag and Friction Drag?* [online], https://airshaper.com/videos/what-is-aerodynamic-drag/DbIIw-WryHY (Accessed 03/01/2024)

Artec 3D Leo (2021) [online], www.artec3d.com/portable-3d-scanners/artec-leo (Accessed 03/01/2024)

Ballinger, A (2018) *Sock Height Rules Will Be Enforced by UCI in 2019*, Cycling Weekly [online], www.cyclingweekly.com/news/racing/sock-height-rules-will-enforced-uci-2019-400212 (Accessed 03/01/2024)

Brownlie, L (2011) *Helping the Swiftest Be Swifter, TED x* [online], www.youtube.com/watch?v=f-bMVS5eYCQ (Accessed 03/01/2024)

Brownlie, L, et al. (2016) *The Use of Vortex Generators to Reduce the Aerodynamic Drag of Athletic Apparel* [online], www.researchgate.net/publication/305080602_The_Use_of_Vortex_Generators_to_Reduce_the_Aerodynamic_Drag_of_Athletic_Apparel (Accessed 03/01/2024)

Croxton, J (2021) *Is the Danish Pursuit Team Using Kinesiology Tape at the Olympics to Circumvent UCI Rules?* Cycling News [online], www.cyclingnews.com/news/is-the-danish-pursuit-team-faking-injury-at-olympics-to-circumvent-uci-rules/ (Accessed 03/01/2024)

D'Auteuil, A, Larose, GL and Zan, SJ (2010) Relevance of Similitude Parameters for Drag Reduction in Sport Aerodynamics. *Procedia Engineering*, 2(2): 2393–2398 [online], www.sciencedirect.com/science/article/pii/S1877705810002596 (Accessed 28/12/2023)

Debraux, P, Grappe, F, Manolova, AV and Bertucci, W (2011) Aerodynamic Drag in Cycling: Methods of Assessment. *Sports Biomechanics*, 10(3): 197–218 [online], www.researchgate.net/publication/51660070_Aerodynamic_drag_in_cycling_Methods_of_assessment (Accessed 03/01/2024)

Hong, S and Asai, T (2021). Aerodynamics of Cycling Skinsuits Focused on the Surface Shape of the Arms. *Applied Sciences*, 11(5): 2200 [online], www.mdpi.com/2076-3417/11/5/2200 (Accessed 03/01/2024)

Kessel, A (2008) Born Slippy, *The Guardian* [online], www.theguardian.com/sport/2008/nov/23/swimming-olympics2008 (Accessed 03/01/2024)

Laser Design (2021) *5 Essential Tools to Master in Artec Studio Software*, LaserDesign [online], www.laserdesign.com/5-essential-tools-to-master-in-artec-studio-software/(Accessed 03/01/2024)

Lewis, R (2021) Interview with A. Mitchell & J. Ledbury, *Silverstone Sports Engineering Hub*, 24 August, Towcester

McMillion, M (2020) Artec Leo helps Vorteq Create the World's Fastest Cycle Skinsuits, *Artec* [online], www.artec3d.com/cases/vorteq-fastest-cycling-skinsuits (Accessed 03/01/2024)

Silverstone Sports Engineering Hub (2021) [online], https://silverstonesportshub.co.uk/about/ (Accessed 03/01/2024)

TotalSim (2017) *Press Release—TotalSim's Rob Lewis Awarded OBE* [online], www.totalsimulation.co.uk/rob-lewis-awarded-obe/ (Accessed 14/09/2021)

TotalSim (2021) *Rob Lewis*, viewed 14 September [online], www.totalsimulation.co.uk/the-team/ (Accessed 03/01/2024)

UCI (2021) *Union Cycliste Internationale, 2021, Regulations*, UCI [online], www.uci.org/regulations/3MyLDDrwJCJJ0BGGOFzOat (Accessed 03/01/2024)

Ultimaker (2022) *Easy 3D Printing, Powered by Software* [online], https://ultimaker.com/ (Accessed 03/01/2024)

Vorteq (2021) *Engineered Sporting Innovation* [online], https://vorteqsports.co.uk/ (Accessed 28/12/2023)

Vorteq (2023) *The Worlds Fastest Skinsuit*, Vorteq [online], https://vorteqsports.co.uk/vorteq/ (Accessed 28/12/2023)

14 Focus Brands Case Study
Implementing a PLM System

14.1 INTRODUCTION

This case study will consider the challenges of implementing a PLM system by Focus Brands, a leading design, sourcing, and distribution company which specialises in sports clothing, footwear, and accessories. The chapter will open by providing an overview of PLM systems, along with details of the company involved in this case study, namely Focus Brands. The chapter will move on to outline the reasons driving the implementation as well as details of the PLM system selected by the company. Insight of the current practice the company has undertaken with its chosen PLM platform Bluecherry Next PLM system, developed by CGS, will be provided along with a SWOT analysis of the strengths, weaknesses, opportunities, and threats of the implementation. The chapter will close by providing useful recommendations to others undertaking a similar implementation process. The key contributors to implementation of the PLM system include Focus Brands employees Timothy Peck (Head of Design), Rebecca Minors (Design Team Manager), and investigating for this case study Jennifer Collinson (Production and Sourcing Manager).

14.2 ABOUT PLM SYSTEMS

PLM is an acronym for Product Lifestyle Management systems which have been around for over 30 years. PLM systems have evolved significantly over time and are now used to monitor the whole life cycle of products and ranges from design to delivery to the end consumer. Benefits include all information being accessible in one location, including design, product development, sourcing, production, and shipping. They can streamline companies' workloads and enable factories to work closer with business by also having access to the PLM system, as PLM systems can be accessed worldwide with access to the internet. Limitations include issues with internet connections such as slow speeds with poor connections and limited information being uploaded due to how the system is written or developed (Conlon, 2020; Harrop, 2022; SAP, 2022; Stark, 2011).

14.3 ABOUT FOCUS BRANDS

Focus Brands is a market leader in design, sourcing, distribution, and promotion for sports fashion and lifestyle brands. The business specialises in clothing, footwear, and accessories, both in the UK and around the world, and is led by a talented team

 DOI: 10.1201/9781003126454-14

of industry professionals with decades of experience in design, marketing, licensing, and distribution. It deals with complex global licensing, subsidiaries, and distributor networks but is also involved in the development of marketing campaigns driven by an in-depth understanding of consumer and lifestyle trends.

Its talented UK-based designers, based in its office in Huddersfield West Yorkshire, bring ideas to life, translating the needs of its customers into original innovative apparel for some of the largest high-street retailers in the world. The marketing team at Focus Brands looks after every element of the marketing mix, including social media, PR, advertising, sponsorships, photography, and graphic design, ensuring to deliver strong and consistent brand presence.

In addition to the UK team, Focus Brands has over 250 members of staff in multiple locations across the world, including the UK, Germany, the Netherlands, and China. In 2022, due to the continued expansion of the business, it became clear that a digital system was needed to link all parts of Focus Brands and give factories easy access to information and data. Prior to this decision all information was transferred via emails and phone calls; however, these methods of communication were proving inadequate in the current fast-paced industry environment that Focus Brands now operates in. Moving to a PLM system was never going to be a quick process and after five years of creating and testing, Focus Brands has fully implemented a PLM system which will replace an abundance of emails, spreadsheet, design files, and tech packs. Putting all this information on the PLM system will also give employees across the Focus Brand's supply chain instant access to live product information.

14.4   PLM SYSTEM IMPLEMENTED BY FOCUS BRANDS

Bluecherry Next PLM system was thoroughly researched and chosen to be the right PLM system for Focus Brands because it links with systems already in place within the business and is also customisable and can be tailored to meet the wider needs of Focus Brands. It is a logical and easy-to-navigate communication tool for both Focus Brands and the factories it works with. The PLM system has become a central hub for communication within the business enhanced by its reporting tools, efficiency on accessing lab dip and sample information, critical paths, and other day-to-day information.

14.5   PROCESS OF PLM IMPLEMENTATION

Focus Brands decided to invest in using the Bluecherry NextPLM system as this linked in with the current ordering and reporting system that Focus Brands already has in place for a commercial and logistics perspective. Initially, the PLM system was introduced to the design, development, and commercial teams. At this early stage, there seemed to be a misconception that the PLM system was an out-of-the-box solution that could essentially *plug and play*. However, after some initial introduction and training sessions with personnel from CGS the PLM platform providers, it was clear that the default set-up of the system was not quite right for the workflow process of Focus Brands. In hindsight, these initial training sessions would have been better replaced with a system development phase, which is where Focus Brands eventually found itself a couple of years later.

In the meantime, a basic functionality of the PLM system was adopted with the design team inputting basic product information and imagery. The purpose of this data input was used purely for populating data into the order-processing system that the commercial team utilises to place orders.

After five years of development and working with CGS personnel in the development of the Focus PLM platform, Focus Brands has arrived at the point where it is now beginning to use the system for much more than basic data input. The process of development over these five years was mainly a combination of internal testing with a small team, then reviewing with CGS over video calls, and exploring options of how to adapt the system to suit the needs of the business and the processes that they follow. This process was incredibly detailed, and it was essential that the individuals involved understood Focus Brands' processes in fine detail, and any developments made to the system were carefully considered from a multi departmental perspective. For this reason, the individuals selected for controlling the development of the PLM system were in-house product design and development staff. The additional workload for these staff meant that they could spend only a limited amount of time each month to work on the system, thus elongating the time to full implementation. This development process involved dealing with a complex web of workloads and detailed development requests, many of which were ad hoc, and this became quite challenging but was overcome through careful and detailed data management.

Once the system was at the stage where Focus Brands deemed it suitable for live use after initial testing, internal training with suppliers was initiated. The design and development teams were split into two groups to ensure the training sessions were not too large. The first session was an introduction which delivered links to a test site, so users could familiarise themselves with the new PLM system. A follow-up session was delivered to run through more detailed processes specific to individuals' job roles. Finally, and most importantly, support and training on the job was facilitated by the team responsible for the development of the system. Key task and step-by-step guides were created to help support frequently asked questions. CGS also created a comprehensive step-by-step guide; however, this became quickly outdated with each update that resulted from more users effectively testing the system, so the smaller specific key task guides proved a little more user-friendly. Supplier training followed a similar process, with an initial introduction over email to the test site along with walkthrough guides. Follow-up video calls were introduced to answer questions and to troubleshoot any problems that arose. It is notable that few problems were initially reported, and all were fed back to CGS where relevant. On reflection, it was found the training that was completed using video calls, although successful, could have been more beneficial face to face; however, with the travel restrictions due to COVID-19 this was not possible.

Looking at where Focus Brands is at this stage in the development and implementation of the PLM system, the following assessment can be made. In terms of user acceptance, the internal and external user base are adopting the system well. What has helped in this positive mindset and approach was to help new users understand how the system will benefit them individually as well as the business. With this overall understanding, where there have been inevitable problems and unexpected anomalies as they push through the initial teething phase, users have been accepting of the imperfections and have been useful in live testing which is helping to further refine and improve the system. As expected, some users have picked the system up quicker than others, and as more users become more familiar with the process, the internal business network of support has become wider. It has also been evident that there were initially some pockets of mild resistance to implementing the system which has now improved as users become more familiar with the system and can see the benefits for the business.

There have been other unforeseen complications along the way, and these anomalies and issues became radically more frequent with the initial influx of users when design, development, and suppliers began using the live system. These errors are slowly becoming fewer. An example of some of the challenges Focus Brands has faced during initial inception included:

- Server performance drops/failures
- Data loss anomalies
- Duplication of data in fabric and trim libraries
- Awaiting CGS to update system

The rest of this section will provide a summary of how Focus Brands is currently using the PLM system from a practical perspective.

The design team members are at the starting point of data input to the system. They begin by creating a detailed CAD of a style in illustrator which is imported into the PLM system, where the style information is built up around it. There is an initial summary of the style, with information relating to the brand, customer, composition, sizing, and colourways. This information is published to the order-processing system where the commercial and logistics teams can access it. The designers also add further detailed information such fabrics and trims. Most of this is selected from libraries that the Focus Brands team has developed within the PLM system. Before Focus Brands had the PLM system, designers would create a detailed annotated CAD drawing. They often had repeated or mismatching information using this approach. The PLM system vastly reduces incorrect information whilst having the ability to fully track every product including the tracking of materials, trims, and resource across the whole product life cycle. This sophisticated level of product management was not achievable prior to the implementation of the PLM system.

Once the design input has been completed, a style is assigned to the development team.

The development team will review the data added by the designer and then assign or customise a size chart which has been created within the system. When the developer is happy that all information required by the factories is within the system, they release the style to the appropriate factory and request lab dips, strike-offs, and samples. The factory will receive a notification that a style has been assigned to them, and they can review and print off the information they require. Once they have submitted a sample a notification comes back to the developer for review.

FIGURE 14.1 Style Fabric Bill of Materials page within BlueCherry Next PLM.

Source: Courtesy of www.focus-brands.com

In terms of the collaborative benefits of the PLM system, there is an area where both the developers and factories can leave comments about a sample that can be accessed regardless of location. Likewise, a full paper trail of comments for every product is also available within the system, which is an invaluable resource enabling continuity of business when colleagues are away or out of the office. The PLM system can also archive data for each product, so if Focus Brands decides to repeat a style, then all the information from the previous development is easily accessible.

The PLM system can also streamline workflows, for example, Focus Brand's lab dips used to be submitted per style and per factory resulting in multiple submissions across the business. Using the PLM system means that only one submission will be required per fabric per factory, and this also ultimately supports the standardisation of colours across ranges.

The PLM-based communication between the development team and factories continues until bulk shipment. Focus Brands is currently building a critical path within the PLM system. This will allow closer monitoring of potential delays. By setting up a workflow, overall monitoring of the product life cycle is possible, and users can be more accountable for what they are working on, meaning most potential delays can be identified early and action taken to avert them. Focus Brands is also building a QC section, which will follow on from the development team's work. At present this is a separate operation, so bringing this into the same programme will hopefully streamline the process.

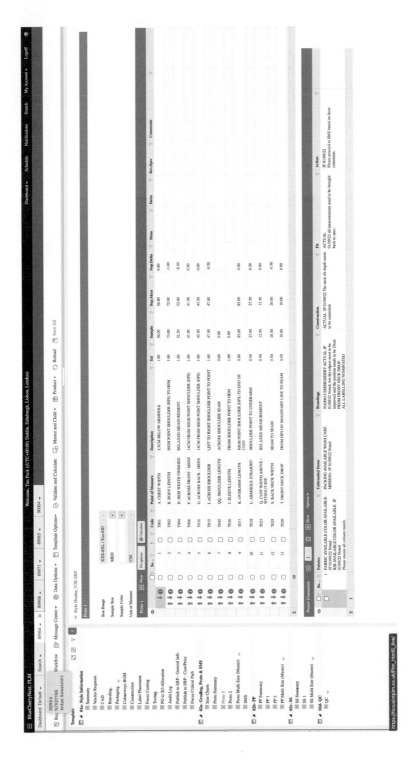

FIGURE 14.2 Prototype Comments page within BlueCherry Next PLM.

Source: Courtesy of www.focus-brands.com

FIGURE 14.3   Lab Dip Summary page within BlueCherry Next PLM.

Source: Courtesy of www.focus-brands.com

TABLE 14.1

SWOT Analysis of the Implementation of BlueCherry Next PLM System

Strengths	Weaknesses
• Track process on workflow	• Time spent to build system, not a dedicated team
• All data in one place	• Potential for user error on data input
• Comprehensive reporting tool	• Slow server speeds
	• Network issues, grind to halt
	• Operating system compatibility issues
Opportunities	**Threats**
• Further development, more departments, customisation	• Other systems becoming obsolete, ordering, and reporting system will always need link with PLM

14.6 CONCLUSIONS

The overall implementation of the PLM system at Focus Brands has been successful to date. This has not been without its faults and problems most of which have been overcome.

The following SWOT analysis shows the key strengths, weaknesses, opportunities, and threats that Focus Brands has and may face.

For others embarking on implementing a PLM system, it can be recommended that continuous and rigorous research and testing from all departments using the system are undertaken. Initial testing of a small collection rather than putting everything onto the system is a good way to measure that all areas are working correctly and help iron out any issues that arise.

Gaining the trust and willingness of the business, its employees, and suppliers to change the way it is working is one of the most difficult yet rewarding parts of the process.

REFERENCES

Conlon, J (2020) *From PLM 1.0 to PLM 2.0: The Evolving Role of Product Lifecycle Management (PLM) in the Textile and Apparel Industries* [online], www-emerald-com.mmu.idm.oclc.org/insight/content/doi/10.1108/JFMM-12-2017-0143/full/html (Accessed 28/12/2023)

Harrop, M (2022) *PLM Report 2022, The Interline* [online], https://theinterline.com/PLM-Report-2022.pdf (Accessed 02/01/2023)

SAP (2022) *What Is Product Lifecycle Management* [online], www.sap.com/uk/products/scm/plm-r-d-engineering/what-is-product-lifecycle-management.html#:~:text=A%20PLM%20software%20system%20is,process%20for%20all%20business%20stakeholders (Accessed 02/01/2023)

Stark, J (2011) *Product Lifecycle Management: 21st Century Paradigm for Product Realisation*, 2nd ed., Springer, London

Index

Page numbers in *italics* indicate a figure and page numbers in **bold** indicate a table on the corresponding page.